THE TARGETING OF MYRON MAY

FLORIDA STATE UNIVERSITY GUNMAN

Assistant DA Pushed Over the Edge

Renee Pittman

The Targeting of Myron May

Copyright © 2015 by Renee Pittman

All rights reserved.

This book is not designed to provide authoritative information with regard to the subject matter covered. This information is given with the understanding that the author is not engaged in rendering legal or professional advice. Since the details of any situation are fact dependent, you should additionally seek the services of a competent professional.

Published in the United States of America

Mother's Love Publishing and Enterprises

ISBN-13: 978-1-7374060-4-4

Dedication

If you bring forth what is within you,
what you bring forth will save you. If you
do not bring forth what is within you
what you do not bring forth will destroy you.

Gospel of Thomas

Table of Contents

Preface   ix

Chapter One: Loss of Hope   1

Chapter Two: In His Own Words   42

Chapter Three: Letter to Congress   59

Chapter Four: "Strange Night of November 19th"   67

Chapter Five: The Renee Pittman M. Connection   103

Chapter Six: Powerful Technology at Play Today   163

Chapter Seven: A Life Discredited and Destroyed   194

Chapter Eight: To Be or Not to Be that is the Question   224

Chapter Nine: NWO Luciferian Agenda And Strategic Distractions   255

About the Author   312

ACKNOWLEDGMENTS

To ALL Victims of the NWO Agenda

PREFACE

These are authentic Myron May "To Do" lists written by May prior to the FSU shooting.

NOTE: I was tempted to correct 'other' to 'others', a simple typo, but do not want to be accused of tampering with any of the documents.

TO DO LIST FOR OTHER WHILE I'M GONE

1. I bought a Springfield 22mm pistol from Bay Guns and Gear in Panama City, FL. Please either pick this gun up or get the money for it.

2. I paid $2,000 to the Arvizu Law Firm in Las Cruces, NM to file a bankruptcy for me on my behalf. According to Arvizu, at least $1800 of this money is refundable. Please get the money back for me.

3. I have a check for $500 from Taunton Family Children's Home. I did not finish the assignment that I was working on, so it's not pressing to get the money back from them. But if they don't mind giving you the check, pick up that $500 check as well.

4. I have a house located at 6302 Grayson Bend Drive, Katy, TX 77494. Please take care of whatever needs to be done with the house.

5. I have several personal belongings at my ex-girlfriend's, Name Deleted, house in Las Cruces, NM at Address Deleted. Her phone number is Name Deleted. Orchestrate that with her.

6. I have several items in storage in Houston, Texas at Public Storage located at 5460 Addicks Satsuma Road, Houston, TX 77084. There is a payment due for these items on December 1, 2014, in the amount of $83.00. Please either make the payment or arrange to have these items picked up from storage. My friend, Name Deleted, lives in Houston, TX. She may be able to help with picking these items up. Her phone number is Phone Number Deleted.

7. I am pretty sure that Name Deleted will want Little Bit. We got this dog when we lived together. Her and her daughter, Name Deleted, love Little Bit as much as I do, and I am sure that they will want to get her and take care of her. Name Deleted phone number is Phone Number Deleted.

8. Gather my personal belongings from the guest house at the Taunton Family Children's Home

BEING A TARGET TO DO LIST

1. FB Msg Renee Pittman M., John Hall (can access through Pittman), Eric Griffin, Herman Winston, Dennis Kucinich

2. Twitter Renee Pittman M. and Dennis Kucinich

3. Post to YouTube (advertising version)

4. Post to YouTube (free version)

5. CNN iReport > Assignments > My Life (under Express Yourself subject line)

6. Post to Vimeo (may have to upgrade to Vimeo Plus)

7. Email Derrick Robinson - derrickcrobinson@gmail.com

8. Email Michael Williams from DASO - michaelwi@donaanacounty.org

Michael Williams is listed as an investigator with the Dona Ana Sheriff's Office

NOTE: Unsure whether the family had the above lists I forwarded them via Mrs. Abigail Taunton of whom Myron May had been staying in the Taunton Family guest house in the days prior to the shooting.

Mrs. Taunton reported to me that she was currently out of town, on December 5, 2014, but that she would insure that May's family would receive the information on Monday, December 8, 2014, when she returned home. Mrs. Taunton confirmed receipt of the two lists.

Renee Pittman

CHAPTER ONE
Loss of Hope

Myron May's authentic suicide letter:

VIA CERTIFIED MAIL, RETURN RECEIPT REQUESTED

November 17, 2014

Dear Ms. Mitchell and Other Addressees:

The purpose of this letter is three-fold. First, I would like to make a sincere plea to you not to let my personal story die. Enclosed within this letter, you will find a USB flash drive containing a personal testimony from me about the financial, emotional, and psychological pain that I have endured over the course of the past few months since I discovered that I was a "targeted individual." I have literally been forced to endure a living hell. There are thousands of targeted individuals within the United States that literally suffer each and every day at the hands of our government. Personally, I have experienced significant harassment from law enforcement in every place that I have been these past few months.

Second, over the coming days and weeks, you will hear numerous people try to label me as a person with mental health issues. If you simply google "targeted individual," "gang stalking," "Freedom from Covert Harassment and Surveillance," or "Dr. Robert Duncan," you will see that what I have experienced, albeit not widely known, is very real. In addition, you can find various videos on YouTube by searching under these exact same search terms. Our government is able to capitalize on this lack of knowledge among the general population to curb sentiments towards questioning the mental health of targeted individuals rather than admitting the truth—that there is a system of covert torture of ordinary, innocent citizens that is happening within our own borders. I have not told a single person exactly what I intend to do—and only you eight people know that I intend to do anything at all—but my goal is to garner some much-needed media attention to the plight of targeted individuals because we are a marginalized group with few financial assets. Coincidentally, that means we get ignored.

Third, enclosed within, you will also find a sample letter to congress. Please encourage as many people as you can to send a copy of this letter to congress. My hope is that if enough people take a genuine concern into the struggles of targeted individuals, then congress will have to do something to stop it once and for all—not like the false machinations of stopping it that took place in the 1970s with COINTELPRO.

I apologize for putting this responsibility on you guys, but you are people that I know and trust. I am confident that Ms. Mitchell will not allow my story to die. I sincerely hope that you will (1) keep an electronic copy of my story for yourself, (2) provide a copy for distribution to media outlets, (3) make sure that Ms. Mitchell gets a copy [although I am sending her a copy, I fear that it may be intercepted], and (4) see to it that if my story is removed from the internet—YouTube, Vimeo, etc.—it will be promptly re-uploaded. I know that I am asking for a lot, but please assist me with this.

Lastly, please whisper a prayer for my soul. I am still a believer, and I honestly feel that there is no hope for me. Consequently, I am making a sacrifice so that others in my same position might have a chance at a normal, harassment-free life. I realize that my methods are not the best selection—and probably will not be perceived as the selection of a Christ-follower—but I have prayed incessantly for months to no avail. There are targeted individuals that have endured this torture for decades without any relief, and what targeted individuals need more than anything is media attention.

OTHER TARGETED INDIVIDUALS

1. Randy Quaid
2. Melinda Fee
3. Stephen Shellen
4. Gloria Naylor
5. Kola Boof
6. Jill Anjuli Hansen
7. Matt Barasch
8. Ted Gunderson (former Senior FBI agent/whistleblower)
9. Jiverly Wong
10. Aaron Alexis

Your brother in Christ,

Myron May

cc:

a) Derrick Robinson
b) Christopher Chestnut
c) Aaron Watson
d) Juan P. Chisholm

e) Chris V. Rey

f) Marc Bozeman

g) Titiana Frausto

h) Rosanne Schneider

i) Joe Paul

On November 20, 2014, nationwide media headlines would read:

FSU shooter Myron May feared 'energy weapon,' heard voices, and thought police were watching him just prior to a rampage at Strozier Library on the Florida State University Campus.

Myron May an FSU alumnus, was killed by police early Thursday morning on November 20, 2014. After shooting five, three suffered injuries. Another of the victims was grazed by a bullet and treated at the scene and released. One, who was shot in the leg, was released from the hospital Friday night. Reported to be critically injured the third victim remained hospitalized and a week after the shooting was reported to be paralyzed.

Myron DeShawn May was a 31-year-old prosecutor with the Dona Ana County, Las Cruces, Mexico, District Attorney's office who had recently returned home to Wewahitchka, Florida, on or about November 6, 2014, after resigning from his position as a Public Defender. Why in matter of just 14 days, would Myron May go into the Florida State University library and take this action?

Prior to this, May left a series of chilling voicemails saying he was being attacked by "Direct Energy Weapons" relentlessly in his chest, and that he had "devised a scheme" to expose the plight of thousands of Targeted Individuals "once and for all."

These chilling messages, along with ten mysterious certified mail packages May mailed to friends were in route and mailed a day before the shooting. May's action appeared to be the culmination of what seemed to be the onset of a mental breakdown in the making several months prior. Noteworthy, this psychological breakdown was after years of highly credible achievements and normal functioning and appearing to happen almost overnight. Just two months ago, in September of 2014, by many accounts, Myron May had a promising career as a young prosecutor.

A mere six months into his position with the District Attorney's office, Myron May would abruptly resign. Later, on November, 21, 2014, a letter would surface which was determined as his suicide letter. In the letter and subsequent letters, Myron May documented that he was being horrifically stalked by unknown assailants and he attributed it to being placed into a government condoned covert program. He stated that the technological harassment had become so pervasive that he could no longer perform his job. Later, a worried ex-girlfriend told police he suffered from mental problems that were getting worse after being admitted to the psyche ward complaining of efforts he felt were impacting and destroying his life.

On Oct. 6, 2014, May's co-workers arrived at work to find May had cleaned out his office and left, leaving a resignation letter on his desk.

The letter thanked DA D'Antonio, was professionally written, and showed no sign of a breakdown or a disorganized thought process. The District Attorney, Mark D'Antonio, later said that neither he nor his colleagues saw any sign of May's mental decline, and stated, "If I had known, we could have gotten him some help. Maybe this wouldn't have happened." He also stated:

"I nearly fell off the chair," a day after the shooting, describing his reaction when he later learned of the shooting. "It was as shocking to me as it could have been. The staff took it hard. He was very well-liked."

This seemed to be the general consensus of everyone who had come into contact with Myron May or knew him personally. For 31 years Myron May had been a highly productive individual and had led a productive, functional life without exception. Family and friends were shocked and horrified. Myron May, had a reputation as a caring, sensitive guy, and self-described over achiever who loved to recite bible verses.

May's first visit to the police station in Las Cruces happened in early September. May stated that "he could constantly hear voices coming through the walls specifically talking about actions he was doing," an officer wrote in an incident report. As an example, May cited a time he had climbed out of a bubble bath and began applying lotion. He specifically stated he heard voices say, "Did you see that, he never puts lotion on," the report states. The Las Cruces police officer said there was nothing they could do officially when May reported this. This was especially true when May responded that he planned to hire a private investigator to find out what was factually happening, who was behind it and why, and that his purpose there that day was because he wanted his report documented for this reason. The fact is that operatives of these efforts can also park vehicles in front of the target's location and use antennas to beam the weapon system into the target's home as shown by an image from one of my other books.

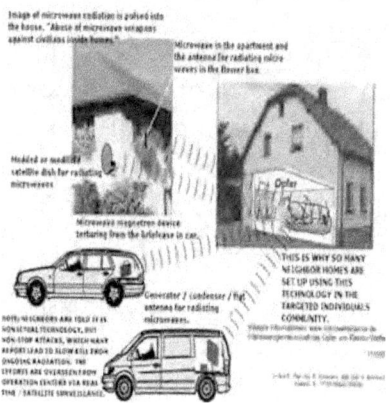

The next night, police were called to the home of May's ex-girlfriend. May had just left, after showing up rambling and giving her a piece of a car, he said was a camera that police had placed in his SUV.

The Targeting of Myron May

May's former girlfriend, a doctor, had dated him for about 15 months before breaking up two weeks before. She reported that she was worried about his state of mind. "Myron has recently developed a severe mental disorder," police wrote in this report. "Myron believes that the police are after him and are bugging his phone and car, as well as placing cameras also in his home."

His ex-girlfriend also stated that he'd been taking prescription medication and had recently been taken to Mesilla Valley Hospital for a mental health evaluation. He had not made any suicidal or homicidal threats, but had been acting erratically. She also reported that likely contributing to Myron May's psychological state was the fact that he had been staying up four to five days straight with no sleep. She was also confused when he abruptly had taken a trip from Las Cruces to Colorado and back again in one day with no reason, the officer wrote.

Police went to May's apartment but he wasn't home. He was not home because they went to the wrong apartment. May lived in 1407. The officers went to 1403. The Las Cruces Police Department then issued an alert for his vehicle for "officer safety and welfare check.

By November 6, 2014, May was home in Wewahitchka now jobless, penniless, and bankrupt in a small city in the Florida Panhandle where he grew up. Wewahitchka, Florida, is a one stop light town, about 30 minutes east of Panama City near Apalachicola National Forest.

Myron May had been a Wewahitchka proud success story and exemplified the American Dream. Very few had made it out of the town and sored to the heights of which Myron May had as an attorney with a several prestigious job to include one in a District Attorney's Office.

After arrival, and a brief visit with his family, May decided that the only choice he had was to attempt to try to pick up the pieces of his life and hopefully start over. He contacted old friends, the Taunton family. He had formed a long-time relationship this family by cross-country running with the son. He explained that he was having issues, and that he needed a place to stay. Ultimately this altruistic family, nationally renowned as charity workers, who run a foster care facility five miles from May's family, offered him refuge in their guest house.

On the morning of November 20, 2014, the Taunton's were among many stunned and who tried to make sense of why May would do what he did. The family reported that May had no history of violence and was a very proud Florida State University Alum. Mr. Taunton also reported that when they heard the news of a shooting at FSU with the gunman from Wewahitchka, he immediately knew it was Myron May. He stated that he knew May had been going through dark times recently which impacted his returned to Wewahitchka, but, he nor anyone else would have even remotely guessed that it would evolved into violence. Myron May, the Taunton's reported, would not hurt a flea and was a gifted young fellow, a sweetheart. The way his life ended was devastating to the family.

A few days before the incident, the Taunton's reported that May had mysteriously disappeared, oddly without telling anyone where he was going. This was either Monday or Tuesday. Concerned Mrs. Taunton texted May asking him was he okay. He texted back saying, "I am alright. Sorry I left without telling you anything. I am just going through spiritual warfare." That was the end of the text and last they heard from May with the rampage occurring Thursday morning just after midnight.

Mr. Taunton stated, during a televised interview, that he now struggled with when and how he could divulge the details of his conversations with May.

Myron May was licensed to practice law in Texas and New Mexico. Now back in Florida, he reported that he was now interested in starting his own firm. The first order of business he told family and friends would be studying to take the Florida State Bar Exam in February 2015.

During the stay, he appeared normal, and May even found time to accompany the family on a recreational outing, was happy and social although still focused and engrossed in studying for the bar exam bringing his law books along.

In my case, when I began to look back for initial contact with May on social networks, I learned that May had begun following me on Twitter on October 16, 2014. However, it was not until November 14, 2014, almost a month later that he would actually began to contact myself, and other people connected to the Targeted Individuals International community of thousands. In fact, it appears that his first real contact was through the Facebook group actually called Targeted Individuals International.

May also reached out to an organization called Freedom from Covert Harassment and Surveillance as early as November 3, 2014. Today this organization, is a gathering and activism resource said to be for thousands of U.S. citizen reporting to be suffering some form of psychophysical warfare and ongoing Directed Energy Weapon attacks in what many realize is a widespread technology testing program combining civilian contractors, high level government agencies, all levels of law enforcement, military technology and military personnel stationed within the United States various at military installations nationwide.

Typical of many targeted individuals after concluding they have not lost their minds, May began to investigate other possibilities for the covert surreal, total destruction of his life of which he could not control, and began to connect to dots. He learned, as thousands have that the trail typically leads to the reality of, and revelation of, a diabolical, heartless program, of which many lives have been totally destroyed in the exact same manner as his for one reason or another. However, very few do not lose, as May had, a high powered and high earning job of $160,000 a year or are destroyed so quickly.

"The Program" is an expertly crafted psychological operation (PSY OPS) combining a mobilized organized stalking network nationwide through community efforts. In fact, the Targeted Individuals Canada website describes Organized Stalking clearly below:

"Organized Stalking is a form of terrorism used against an individual in a malicious attempt to reduce the quality of a person's life so they will: have a nervous break-down, become incarcerated, institutionalized, experience constant mental, emotional, or physical pain, become homeless, and/or commit suicide. This is done using well-orchestrated accusations, lies, rumors, bogus investigations, setups, framings, intimidation, overt or covert threats, vandalism, thefts, sabotage, torture, humiliation, emotional terror and general harassment. It is a "ganging up" by members of the community who follow an organizer and participate in a systematic "terrorizing" of an individual."

– Mark M. Rich

TI - Targeted Individual - is a person being targeted with OSEH by a group of individuals called "perp" for the purpose of human experimentations.

OSEH - Organized Stalking Electronic Harassment - are methods use by perps in targeting a specific person for the purpose of inducing harm and possibly death.

DEW - Direct Energy Weapon - are device used for OSEH purposes, weapons can be microwave with pulp frequencies, v2k or other electronic and hearing devices.

V2K - voice to skull device - is a weapon use for transmitting voices with low or high frequencies. Voices can be for commands or harassments attacks that may look like the TI's own voice. V2K can also use to induce or manipulate dreams or to deprive TI sleeps.

www.targetedindividualscanada.wordpress.com

These professionally organized campaigns consist of everyone from, your local grocer, a gardener, janitors, doctors, teacher, ex-cons, and average Americans, to include, off duty fire and police department personnel. Another facet is the InfraGard which connects local business owners to this program. Covert stalking and terrorism are also organized within Neighborhood Watch Programs and Community Oriented Policing Group programs. The terrorism entails Gas lighting, relentless technological harassment, noise campaigns, damage to property, etc. It is typically backed up and overseen by numerous state-of-the art operation centers monitoring and tracking the target, for situational awareness, and these operations factually are using various types of Directed Energy Weapons used for coercive psychophysical torture.

However, through media and disinformation, and gag orders associated with National Security Letters, the public's perception of targeted individuals has been carefully crafted. As a result, Targeted Individuals are people who are often perceived as conspiratorial or delusional who contend they are targets of spying, harassment or abuse, by electromagnetic radiation weaponry in delusions of grandeur. The fact is today, it is the average man and woman being used to test the technology used in these programs. There is disbelief because it is very difficult to prove because the technology is deployed as radio frequencies, unseen to the human eye and detectable.

On, Friday, NBC News reported that May had reached out to another "targeted individual." I also would like to point out that I hope the public is not left with even a remote possibility that I would ever go postal. In my case, this is what pen and paper and book publications are for as with many others using various other exposure measures.

When I was later asked what my take on May's mental state when we spoke, I reported that he told me calmly, that "He just didn't want to go on living like this." However, to me, it was said without any sign of duress, stress, or anger detectable in his voice. I again reiterated that on a deeper level, I could not determine the state of someone's mind from a person; I had never met personally, and had only known for a mere six days, via social network. I also pointed out that based on reports from those that did know May personally; he showed no sign of extreme attitude of deadly intent or suicidal ideation.

The evening before the shooting, May would leave three chilling voicemails on my cellular phone, between 7:19 p.m., 7:22 p.m. and 7:42 p.m., CST, Wednesday evening approximately five hours before the shooting at the FSU library. I live in California so this was one hour ahead.

"I am currently being cooked in my chair. I devised a scheme where I was going to expose this once and for all and I really need you." He also said. "I do not want to die in vain" in one of the messages.

Myron May made these comments of which I would later forward via email to the NBC News reporter covering the case. The result was a news report and printing of a portion of the voicemails. This was after the voicemails were first authenticated by a relative of Myron May as factually being May's voice.

I would not listen to the messages until morning around 7:30 a.m., on November 20, 2014. By that time, Myron May was long gone.

In an email sent at 8:19 p.m. PST, 9:19, CST, Myron May wrote: "I've been getting hit with the "direct" energy weapon in my chest all evening. It hurts really bad right now." Sadly, many targets get the directed energy weapon psychophysical torture and we endure and carry on. As I have documented in "Covert Technological Murder" due to book publications and the inability to control me, a different strategic effort has evolved in my life. It is known as "slow kill." This is accomplished by focus on my joints with the beam set in slow cook mode. Because, many, many people endure, I never expected May to take the drastic action he did.

In hindsight, it appears that May, based on evidence he subsequently sent by gmail to me revealed that he was definitely in a powerful spiritual warfare long before the voicemails. His contact with the targeted individual community or whatever he reportedly told the Taunton family could reveal this. This fact is revealed by his well thought out instructions and directions of how to handle his affairs, his belongings, pet, and the request to save license plates of stalkers, as evidence date back as far back as October 21, 2014. He also had a desire to make YouTube and Vimeo videos of which he hoped to download or did download, but were taken down, along with the flash drive he reported as part of the certified mailings.

After the fact, it appears he obviously lost all hope, likely recognized after researching this program, then connecting with others within the targeted individual community, and the realization that once in this program, no one has never, and I do mean ever, been released from it. Obviously, the chance cannot be taken that an unmonitored person could expose its reality. We are talking about hired individuals working in shifts around the clock and of who maintain their positions up till retirement.

With this reality, glaringly a reality for targeted individuals, May was likely now in unbearable emotional pain, shame of demotion, and more importantly labelled mentally ill which is 100% career destroying in and of itself. The Wewahitchka Star Child had been reduced to absolutely nothing and this documented sensitive spirit it appears, simply could not bear it.

One of the certified mail packages arrived in Texas, according to the Associated Press. Another was intercepted by postal authorities in Orlando. Joe Paul, the last person listed as a recipient on the suicide letter, who attended Florida State University with May, alerted Tallahassee police about the package headed his way immediately. When Joe Paul was questioned, he stated his curiosity along with each recipient's similar curiosity of what the package held.

The first two victims were shot just outside the library. May then entered the library and shot the third but did not pass through the lobby turnstiles. When he went back outside, he just stood, and then was confronted by police.

On Friday, Leon County State Attorney Willie Meggs said he'd been told May had shot at six FSU and Tallahassee police officers who first arrived on the scene. A day earlier, Tallahassee police Chief Michael DeLeo declined to answer questions about whether May had fired or pointed his weapon at police. The first reports reported that May had simply refused to relinquish the weapon before law enforcement opened fire in hail of bullets. The second and subsequent reports stated that May took a shot at the officers.

Because the Postal Inspector, and FBI confiscated and intercepted the certified mailing sent by Myron May out from Tallahassee on Tuesday or Wednesday, a copy of the letter, actually dated November 17, 2014 determined to be May's suicide letter was sent to me, at my request by Derrick Robinson, Director, Freedom from Covert Harassment & Surveillance on the afternoon of November 21, 2014. This was after the official team comprised of Postal Inspector, someone identified as an FBI agent, and a local female sheriff left my home around 10:30 a.m., Friday, November 21.

Badge example only

The Targeting of Myron May

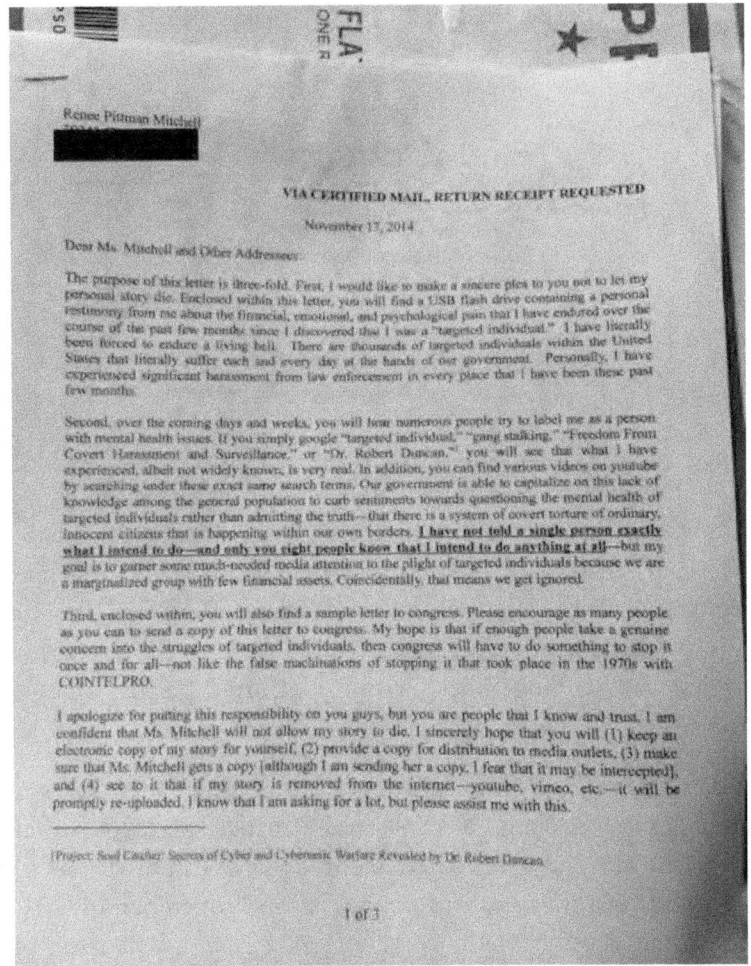

Included also with the image forwarded to me by Mr. Robinson was an image of the flash drive disk. Myron May had told many recipients that it would also be inside the mailing as shown below in an image also provided by Robinson.

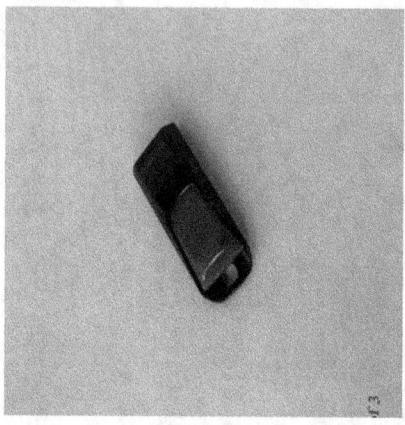

After the trio left, I immediately called an associate in North California. In fact, it was she of whom I had been on the line with when May began the succession of voicemails the evening of November 19, 2014, of which I did not respond. We were trying to figure out what was up with him especially due to the secrecy of his saying he was sending videos coming from a perfect stranger to strangers.

Earlier that day that May had sent confirmation of mailings to me for credibility of mailing. When I realized that she too had been in contact with May, and was listed as a recipient, I immediately phone her. She told me that she was contacted on November 14, 2014 for the first time via Facebook by May. As I stated, when my phone continued to beep as call waiting, we were discussing whether he was credible or not. I remember asking her, "Dang, what is up with this guy" with a stranger calling so repetitively. One of the main reasons that I began to suspect May as misrepresenting himself was after he sent images of the postal receipts. The postal receipts did not look official to me. I felt he was an imposter in a long line of individuals intentionally put around me to silence and entrap me. I just did not know for sure and did not want to get involved at all whether true or not. I, continue to fight my personal battle for my on a daily basis.

Needless this woman and I we both were in shock, and I was floored when she phoned me after work on November 20, 2014 around 3:30 p.m. I had been working on my small inspirational book, "The Heart is Another Name for God" turning it into an audio book and had not turned the television on. She asked first how I was doing, I responded nonchalantly, "fine" not having a clue. She then broke the news. I was literally brought to my knees. I simply could not believe it nor could I believe that something like this had happened so close to me.

Later many would question why I willingly gave permission for confiscation and interception of the certified mailing and handed the package over without a second thought to the Postal Inspector. This was due to the details of the shooting pouring in and the reports stating that three people had been injured.

The fact is, after learning the details; the rampage appeared to be "suicide by cop." I did not want to be a part of what happened, or connected to it in any way.

People with less in their lives are hanging on and fighting a good fight. Again, I just never expected that a highly education attorney would have done this, technologically nudge into it or not. However, his losses must have been a great and intensified sadness for him. Although this does not by any means make the choice he made right.

The objective when a person is placed into "The Program" is jail, a lengthy stay in the psyche ward, suicide or being pushed and pushed to reaching a breaking point.

With a knot in my stomach, I was greatly disappointed in the direction May had chosen as his only option. However, many targeted individuals not only within the United States, but globally, know that things are not what they seem or as reported in the media related to these situations and that there is a much deeper dynamic literally at play in these operations and that the story is not being completely told.

I also could not shake the feeling that somehow Myron May's focus solely on me as he grew closer and closer to his "scheme" that something unspoken just did not feel right to me instinctually. The fact is had I answered the call waiting requests, I possibly could have been, by appearance, unwillingly involved myself and likely deemed guilty by association.

On November 30, 2014, as the Drone motivated energy weapon attacks continued to escalate around me, snapping me back to reality, I got out of bed and decided to add the Myron May story, the shooting and his connection to me, to court documents associated with my open Pro Se civil complaint case in the Central District Court of California. I began searching through the texts, and emails sent to me from May for anything that I could use as an Exhibit for the judge to show another life, targeted was destroyed in "The Program." I located the mail from the Gmail address: Commu Nicate addressed to me by May sent also on the November 19th, right before the voicemails and the FSU tragedy. He had also sent another email to my Yahoo email account, showing postal mailing receipts. While taking an in depth look at the attachments of this Gmail, I was surprised to find several of the items in this book which were more than likely part of the certified, confiscated mail packages.

As shown below, I later found as a large number of images. There turned out to be 34 attachments, although excluding the videos and of course flash drive.

Note the hyperlinks have been disabled by me.

On Wednesday, November 19, 2014 7:18 PM, Commu Nicate <comm.you.nick.8@gmail.com> sent:

20141015_140529.jpg (Proof of certified mailing)

Strange Night of November 19th.odt (Email to Renee Pittman M. detailing screen shots of proof of certified mailing)

Screenshot_2014-11-19-20-18-08.png (Proof of mailing)

The Targeting of Myron May

Screenshot_2014-11-19-19-03-05.png (Proof of mailing)

Screenshot_2014-11-19-19-03-16.png (Proof of mailing)

Screenshot_2014-11-19-19-02-55.png (Proof of mailing)

Screenshot_2014-11-19-19-02-43.png (Proof of mailing)

Screenshot_2014-11-19-17-56-40.png (Proof of mailing)

Screenshot_2014-11-19-17-56-26.png (Proof of mailing)

Screenshot_2014-11-19-17-56-01.png (Proof of mailing)

When I looked in detail, I found several letters which were likely also part of the certified packages.

Loose Items.odt (This is the "To Do" list May created for his family on what to do with his belongings)

Letter to Renee and Others.odt (This is the identical suicide letter FFCHS received forwarded by the Postal Inspector to the Director)

Letter to Congress.odt (This is May's Draft Letter to Congress he hoped targets will send out nationwide)

Copy and Paste of Email for Preservation of Evidence.odt (This is a letter dated, October 21, 2014, to a person he knew to preserve Iphone images of license plates of stalker saved in the phone assigned to him by employer)

Being a Target.odt (This is Myron May's story in detail, detailing his targeting, how and why and what pushed him over the edge)

Being a Target To Do List.odt (This is another "To Do" to contact, Dennis Kucinich, myself, & others, along with creating youtube and vimeo videos)

From this point on are images of numerous license plates May said had been following him with several attempts to run him off the road. There is also an image of a postal employee when he was mailing the certified packages at the Post Office, and an image of someone he said followed him into the post office in his ongoing organized stalking and covert harassment, and an image of a person whom he reported to police had brandished a weapon while sitting in a pick-up truck watching as he packed the U-Haul to leave Las Cruces, New Mexico enroute to Florida.

20141119_181601.jpg
20141119_181521.jpg
20141119_181450.jpg
20141119_181344.jpg
20141119_181202.jpg
20141119_181113.jpg
20141119_181037.jpg
20141119_181006.jpg
20141119_180929.jpg
20141119_180810.jpg
20141119_180735.jpg
20141119_180707.jpg
20141119_180607.jpg
20141119_180529.jpg
20141119_174724.jpg
20141119_150247.jpg
20141119_150239.jpg
20141119_150230.jpg
20141119_150224.jpg
20141119_150218.jpg
20141119_150208.jpg
20141119_150159.jpg
20141119_150149.jpg
20141119_150140.jpg
20141119_150133.jpg
20141027_153242.jpg

20141027_153226.jpg
20141022_090429.jpg
20141021_224231.jpg
20141021_224221.jpg
20141021_085857.jpg
20141021_085856.jpg
20141021_085855.jpg
20141021_085746.jpg
20141021_085746 (0).jpg
20141021_085744.jpg
20141021_084850.jpg
20141021_084848.jpg
20141021_084847.jpg
20141021_084844.jpg
20141021_084843.jpg
20141021_084831.jpg
20141021_084830.jpg
20141021_084828.jpg
20141021_084826.jpg
20141020_163357.jpg
20141015_133953.jpg
20141015_135121.jpg
20141015_132511.jpg
20141015_125417.jpg
20141015_110213.jpg
20141015_080033.jpg

After reviewing everything, one thing is certain, May's letters and all of the information he wrote are clear, concise, well organized, and reveals an organized thought process. This is in spite of, as reported by many, during a difficult time for him and being under a tremendous strain. This is also in spite of what law enforcement reported to have confiscated from his belongings as material having been written by a paranoid schizophrenic, and the information written as the unintelligible ramblings. One thing is certain, if everything occurred as he wrote, he was 100% a Targeted Individual. His story is the story of thousands of others and identical to many.

When a person is in a weakened state they become pliable to programming. Many familiar with "The Program" substantiate its existence via official, unclassified documents, and highly credible, patents, etc., and documented historical facts. Too many May was the obvious victim of some type of mind control program and used as a patsy. The fact is there are documented electromagnetic brainwave frequencies which can control and alter a person perceptions and consciousness.

The Brain Wave Frequency Series - Mind Control

"There are four types of brain waves: alpha, beta, delta, and theta. The four basic models of Monarch slaves have the same names as these four types of brain waves.

High level Illuminati models may have programming that includes all of these types. According to one ex (?)-government source, the CIA has been labeling their harmonic-created total Mind-controlled slaves by the following:

Bravo 2 series models are men programmed to run the Beast computers as shown by the link below shown is an excerpt from the book entitled Project L.U.C.I.D.

Project L.U.C.I.D. - The Beast 666 Universal Human Control System, by Texe Marrs, 1996

Here is the book's Introduction:

Content of Project L.U.C.I.D. ", pg. ix - xix, Introduction

"We Knew It Was Coming

The Beast 666 Universal Human Control System has been designed and is being implemented in America and throughout the world. We knew it was coming. Now it's here, and soon, there will be no place left to hide. By the year 2000, Big Brother's evil, octopus-like tentacles will squeeze every ounce of lifeblood out of the people. A nightmarish, totalitarian police state is at hand.

That is the thoroughly documented message-and momentous warning-sounded in this book. Do not for an instant think that you and your loved ones can escape the monstrous behemoth which lies in our path. Once Project L.U.C.I.D. is fully operational, every man, woman, and child will fall under the power of its hideous, cyber electronic grasp.

Another excerpt states:

The number 666 will signify the hellish master responsible for the devilish invention of this Universal Biometrics Card and its interlocked, computer network.

And,

L.U.C.I.D's cyberspace dossiers will be made instantly available to thousands of probing, faceless, police state agents and bureaucrats around the globe. America's CIA, IRS, and FBI, the Russian KGB, Europe's Interpol and Europol, Israel's Mossad, Britain's military intelligence-these are just a few of the planetary-wide law enforcement, military, and intelligence agencies having access to your dossier and to mine…"

The Delta programmed series are models for espionage and assassination. Juliet series are sexual mind-controlled slaves.

Kilo 5 series is military espionage. Michael 1 [Michael Dunn] series slaves are CIA agents under total mind-control. Much of the high-level programming in the 1980s and 1990s is no longer done with human programmers, but is done via programmed machines using drugs, electricity and harmonics…

Delta programming is military-assassin programming that has trickled into popular consciousness through movies like La Femme

Nikita, its American remake, Point of No Return, and The Long Kiss Goodnight."

It seems as though various agencies working with clandestine CIA/MK ULTRA/Monarch mind control programs may be assigned the assorted brain waves frequencies to work with. Each agency has its own people and can be identified by signature frequencies, Alpha, Beta, Theta and Delta.

The Delta Series-Mind Control

MK DELTA, and its associated program MK ULTRA were on the surface essentially secret mind control and torture/interrogation operations run by the CIA. Both MK ULTRA and MK DELTA involved the surreptitious use of LSD and other bio-chemicals in clandestine operations. However, MK DELTA was actually related to the clandestine mind control and (JFK) assassination operations of a high-level military intelligence, anti-communist-ultra conservative and an extreme racist, Dr. Alton Ochsner. He was centered in New Orleans at Tulane University Medical School, and the Tulane's DELTA Regional Primate Center.

Many targeted are being tested on at various levels within this massive, global technology testing program designed from ongoing research, TESTING, and development programs dating back decades.

In Myron May's case it appears that he was jumped up to one of the highest levels of manipulation, after emotional defeat and destruction of his life weakened him. The losses in his life a form of trauma, then determined possibly useful, strategically, leaving him open for Manchurian programming out of his sheer despair and to be used for NWO goals. Many who are awake appear to be reporting a connection to globalization as shown in the article link below:

LINK to article:

http://www.equestrianempire.com/the-conspiracy-behind-fsu-shooter-myron-may/

May in the suicide letter lists several, well publicized and well-known targeted individuals who are on the record stating they are targeted in "The Program" and who have become activist and whistleblowers bravely speaking out regardless of the stigma and of who have also suffered great loses. At the top of the list is Randy Quaid.

Excerpted below, is an article by whistleblower and activist, Deborah Dupre, examiner.com, entitled:

Randy Quaid blows whistle on crime against celebrity Targeted Individuals, dated, October 31, 2010

Randy Quaid

Randy Quaid speaks out about Targeted Individual phenomenon

In a Vancouver press conference, actor Randy Quaid described part of the typical plight of a Targeted Individual; naming eight close Hollywood associates he believes were targeted and killed. The targeting system is highly organized; involving people at all levels of society including in banks, the Department of Justice, and even family members who are often co-opted to help destroy the innocent target, as Quaid has learned first-hand.

From a written statement, Cousin Eddie in the National Lampoon's Vacation movies and Brokeback Mountain star, Quaid explained the same organized criminal activities hundreds of Targeted Individuals have reported to Dupré. Each of these activities is designed to ruin the targets' careers, their finances, and their well-being - and thus silence them.

Such targeting can and often does last decades. Mr. and Mrs. Quaid have been targeted twenty years according to his statement on October 29...

Thursday, April 05, 2012

Randy Quaid Is Not Stalking - There Are Many Including Actor Stephen Shellen the Only Hollywood Actor Being Targeted for Organized Stalking and beamed high-tech assaults.

Stephen Shellen

Actor Stephen Shellen Describes His Experience

"I've been stolen from, gas lit, blacklisted, surveil lanced and death threatened since 1995.

Melinda Fee

Los Angeles Support Group Meeting, Sunday, August 18, 2013

Melinda Fee, a retired Hollywood actress was the guest speaker at the above support group meeting. She spoke on issues related to her personal experiences as a targeted individual by her experiences of organized stalking, electronic harassment and some chemicals. It is well-known that many in Hollywood are also targeted.

Gloria Naylor

Gloria Naylor is a famous writer of six novels, particularly "The Women of Brewster Place". A native of New York, Naylor graduated from Brooklyn College in 1981. She later received a Master's degree at Yale University. She received the Guggenheim Fellowship Award and the National Book Award in 1983 for her first novel, The Women of Brewster Place. Brewster Place was turned into a television mini-series by Oprah Winfrey. She became a targeted individual in 1996. Her experiences started with constant surveillance (being followed everywhere), organized harassment (wire-tapped, computer tapped, noise campaign etc.) and ultimately assault by electronic weaponry (technological hearing voices effect). Because Ms. Naylor was living comfortably, she managed to maintain her financial situation, unlike many targeted individuals, who are destroyed and become unemployed and homeless. However, she experienced depression which she sought help and was prescribed Haldol. Haldol is the typical antipsychotic drug used to treat schizophrenia and psychotic state/delirium. In one of her petition letters, Ms. Naylor declared that this experience denied her civil and human rights. She studied the phenomenon and decided to do something about it by writing the novel.

1996 detailed NSA covert attacks on her person and is described below on amazon:

In 1996 novelist Naylor moved to secluded St. Helena Island. A minor fracas with a crotchety neighbor, whose brother worked for the National Security Agency, set into motion a series of events that made Naylor the object of close scrutiny by the government. The intensely private Naylor found herself in the company of other prominent black writers and activists targeted by the government. She was harassed until she feared for her sanity and consulted a psychiatrist. When she discovered that her tormentors were using mind-control techniques, she fought back with the only weapon she had--her writing talent. Naylor's account, with first-person recollections of her experience and third-person speculation on the motives of her tormenters, not only raises alarms about government actions but also raises many questions about this work. She includes addenda with research and litigation regarding government experiments with mind control. Whether readers think Naylor has been the victim of pernicious abuse or that she has suffered a breakdown, her incredible book will spark debate about government surveillance and the blurring of the lines between fiction and nonfiction.

In my opinion, one of the worse things a target can do is take Nazi psychiatric medications. Unless you have a degree of credibility prior to the targeting, which backs up a normal and functioning life prior to the targeting this can work against. Taking anti-psychotic leaved the greatest stigma via documentation that target accepted mental illness and agreed they were mentally and had to be medicated. Secondly, medications cannot stop the "Voice of God" technological effects.

Because of this target can become unwitting human test subjects another major area of mind control which is the testing various anti-psychotic medication and/or increased dosages of current prescriptions. However, if the target has a degree of credibility, and an avenue to expose the reality of a factual technological psychological electronic program, and its very real effects, and can detail its legitimacy, the credibility can over shadow the stigma.

An excellent video to watch and to grasp the reality of psychiatry'

"Don't Miss This One "Psychiatry and Industry of Death (Full Version)

LINK: https://www.youtube.com/watch?v=gvdBSSUviys

Working hand in hand this program the psychiatric community has continued playing a prominent role in the advancement of Behavior Modification studies which includes "Remote Neural Monitoring."

Googled "Targeted Individual" and if you do, you will find a myriad of powerful, substantiating information, detailing ongoing programs, again operating at even local police levels, along with ongoing efforts of activist and whistleblowers, from every walk of like, determined to expose the reality of what is factually happening today and also how lives are being, systematically, covertly, technologically destroyed. Also, thousands report, this is after successful discrediting of mental illnesses.

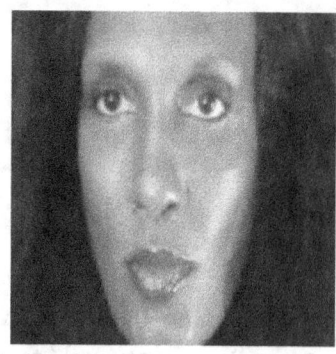

Kola Boof

There is a video out that suggests that Kola Boof is a mind control victim who is under the control of some intelligence agency. Kola Boof appears near the end of the video.

The video can be seen at the following link:

Some believe that the intelligence community provides sly hints of Kola Boof's true nature and origins. Shown below is the cover of Kola Boof's autobiography

The monarch butterfly tattoo is the code/symbol of the CIA's Operation Monarch, which was/is one of the major mind control operations under the CIA's MK-ULTRA program.

The butterfly tattoo is used to identify mind-controlled individuals to operatives in the field. According to mind control investigators, use of the monarch butterfly tattoo code went out of favor after its use was exposed in the general press.

Surfer Model Jill Anjuli Hansen

It was after Jill Anjuli Hansen's TED (Technology, Entertainment, Design) speech, "Open Mind, Open Heart" September 2010, at Fullerton College, that Hansen reports becoming a targeted individual or as she puts it, the microchip within her was activated.

Some would argue that many of the experiences she details happening to her in her life, during this speech, are classic psychological manipulation and "Voice of God" technology

Link: www.youtube.com/watch?v=-mzvAAuCo1c

By May of 2014 Hansen, now penniless was in jail and expecting 10 years in prison after vehicular assault for trying to steal a 73-year-old woman's car and the headlines read:

Surfer and model' accused of attempted murder of 73-year-old woman smirks through first court appearance...

Jill Hansen, 30, pleaded not guilty to attempted murder Tuesday

She is alleged to have run down Elizabeth Conklin, 73, in the garage of an apartment complex in Waikiki

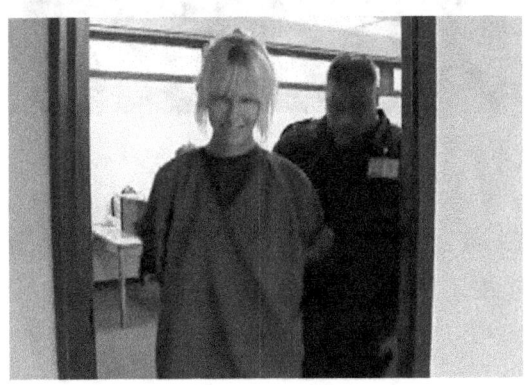

In custody Hansen looked unperturbed as she made her way out of the courtroom, flashing a menacing grin at waiting camera

The court would later order a full psychiatric evaluation.

One year prior, Hansen was a special guest on Freedom From Covert Harassment and Surveillance group conference call, February 23, 2013, Episode 272, during which time she detailed her plight, how and why.

She detailed the spiraling down of her life after the TED speech went viral, and the repercussion of her father who she stated as being heavily involved with the Illuminati, as a high-ranking IRS official.

Hansen stated that both her father and mother sought to profit, exploit, and control her since birth motivated from a child through a strong Illuminati allegiance. In order to do so she was implanted tracked, victimized by the technological hearing voices effect, a victim of organized stalking efforts and her life totally destroyed.

Matt Barasch

Matt Barasch, Attorney, business owner, appeared on an episode of Dr. Phil entitled: I Swear I'm Being Cyber Stalked, Wiretapped & Followed... Myron May listed Matt Barasch due to his appearance on the Dr. Phil show.

Many do not understand the dynamic of the decade's long mind control effort or advanced covert technological usage of technology designed for mass population control. You can be sure of one thing, anyone wanting to keep their job in television will not report on the factual reality of what is happening today. This includes television personalities with a large audience such as Dr. Phil. A media gag order is also likely playing a major role. The truth could reach millions and result in national disgrace.

After decades of technology testing on human subjects, today there appears to be a Standard Operating Procedure (SOP) for "The Program." Illusion and perception management are pivotal and key for discrediting and one the very first objectives setting the stage.

The Targeting of Myron May

For example, the target's attention is drawn to ongoing street theater, while suggestions are beamed at the target through patented consciousness altering and subliminal suggestion electromagnetic radio waves. Understand this is played out along with factual, formed, and combined mobilized organized community stalking efforts by an agency. The target, once held in this psychological altered state, after a factual incident, becomes convinced that each and every incident is about him or her as operation center personnel hope. In reality, at times, it is someone(s) sitting at a computer terminal pushing the target's buttons, biometrically, toying with the target and using others around the target who have not a clue why they said or did something innocently. I have witnessed this personally many, many times as documented in another book. Any and every one can become a pawn to include those inside the target's home. For example, Myron May described that when at the grocery store he had a powerful overwhelming feeling, likely through radio frequencies beamed at him, that other patrons were watching him and going to hurt him. He stated that the feeling was so intense emotionally that he called a friend, a policeman, to escort him home. Understand that first and far most, this is a non-consensual human guinea pig testing program and it is open season on everyone.

The capability to create recorded sound via satellites, in my case, is exampled by a recent incident of ringing of my door bell and loud pounding on my door with no one there as operation center searched for useful fear technologically scanning me. This is a highly effective method used, in many ways, to discredit the target when reporting a phantom person, after an initial real incident, that also rings doorbells, in my case, and pounds loudly. Remember when the trio knocked (FBI, USPS inspector and local female sheriff) the day after the Myron May incident for the Certified package sent to me, they also rang the doorbell and pounded loudly which was identical in sound and it appears likely recorded. I later recognized this as a hope to set the stage for intimidation. Many reports and typically, in all cases, having called police when these incidents happen and it is then documented that they are having a psychological crisis and no police report is ever filed. In reality it is secret police units today spearheading these covert Spy Ops at the federal, state and local level of law enforcement who are also mobilizing community stalking efforts using disinformation about the target. As with many others, prior to the targeting, thousands have reputations of being stable, completely normal, and functional. Understand also that the Spy Ops game is designed to create a reality of something, witnessed by the target initially as factual, then played out repeatedly for mind control.

No longer are human guinea pigs, mind control, testing reserved for the poor, African Americans, always the first stop historically, the homeless, those confined in psyche wards, or jailed. The intent has materialized for further testing across the board and at different levels of education.

The fact is non-consensual human experimentation has been going on for decades initially via the Signal Intelligence (SIGNIT) Electromagnetic Frequency (EMF) Scanning Network. The program is documented to use brain scanners deployed via signal intelligence satellites and HAARP, although we were told that HAARP activities were officially closed down likely after its mission was accomplished.

Using electromagnetic radio frequencies deployed by these two and other advanced technological means, to include, again, drones, over the horizon, radar can remotely scan the brainwaves of any U.S. citizen for tracking, technological harassment, and deployment of a bio-coded weapon aka Directed Energy Weapon system of which May reported being hit. He was being scared to death or scared into bitter action.

Mind control programs which use EEG Heterodyning technology to synchronize Artificial Intelligence computers, with unique the brainwave print of each American citizen, thus turns the electromagnetic spectrum into an invisible mind control prison. Again, the technology allows the biometric signature of everyone to be downloaded and recorded into supercomputers of which military personnel such as the United States Air Force are now operating and using in connection with civilian law enforcement agencies, and Department of Homeland Security.

Former FBI Agent Ted Gunderson

Former Special Agent Ted Gunderson suspected he would be "taken out" eventually. As a whistleblower disclosing crimes of the highest order, Gunderson attested to suffering endless harassment and attempts on his life, from operatives entering his home to sneak poisonous liquids into the wall heater, to phone tapping, personal computer hacking, and years of surveillance by "groups and individuals" in ground vehicles, helicopters, and on foot. Agents of his undoing were everywhere. Law enforcement were worse than helpless…they were complicit. Excerpt, Ted Gunderson: Death of a Public Paranoid, by Doug Mesner, "The Process" July 31, 2012.

"The late great Ted Gunderson was a former FBI chief who spent decades of the latter half of his life investigating the New World Order and Satanism. His inquiries led him to discover a common thread running through the global conspiracy: a worldwide child sex trafficking and pedophilia ring that sold children into slavery and flew them to places like Washington DC to be used in sex orgies by politicians. Gunderson found that, in America, more than 100,000 kids were going missing every year, and that the FBI, CPS and White House were fully complicit in the coverup." Excerpt Makia Freeman. Freeman is the editor of The Freedom Articles and senior researcher at ToolsForFreedom.com, writing on many aspects of truth and freedom, from exposing aspects of the global conspiracy to suggesting solutions for how humanity can create a new system of peace and abundance.

These situations could create trauma-based children who then would be successfully programmed as Monarch Manchurians, also known as Delta Programmed, or other levels, i.e., Alpha programmed, sex slaves, etc., then becoming useful for a variety of purposes.

Monarch programming historically is the trauma-based technique documented as part of MK ULTRA heinous scientific studies using human guinea pigs which includes children. The program lasted, officially, for well over 20 years (1950-1974) before exposure and before, it appears, going underground.

Note also the strategic successful connection of mental illness to whistleblowers.

Apr 8, 2009 9:24 AM CDT

Jiverly Wong

The Targeting of Myron May

Binghamton shootings

The Binghamton shootings took place on Friday, April 3, 2009, at the American Civic Association immigration center in Binghamton, New York, United States. At approximately 10:30 a.m. EDT, Jiverly Antares Wong, a naturalized American citizen from Vietnam, entered the facility and shot numerous people inside. Wong had taken English language classes there from January to March 2009 before dropping out.

By many accounts, of those aware today, the Targeted Individual symptoms was clearly visible by several incidences and evidence and also glaringly in the letter Wong wrote describing the harassment he received from the people he described as undercover cops.

Aaron Alexis

The Washington Navy Yard shooting occurred on September 16, 2013, when lone gunman Aaron Alexis fatally shot twelve people and injured three others in a mass shooting at the headquarters of the Naval Sea Systems Command (NAVSEA) inside the Washington Navy Yard in Southeast Washington, D.C. The attack, which took place in the

Navy Yard's Building 197, began around 8:20 a.m. EDT and ended when Alexis was killed by police around 9:20 a.m. EDT.

It was the second-deadliest mass murder on a U.S. military base, second only to the Foot Hood Shooting in November 2009.

Aaron prior to the incident also reached out to Freedom from Covert Harassment and Surveillance and also reported signs of typical stalking and technological harassment.

Targeted Individuals regularly report on how, not only themselves, but other members of their families, friends, and even people with whom they are in some way acquainted, or are associated suddenly end up dying of rapidly spreading forms of cancer, heart attacks, aneurysms, created by Sonic Weapons, or Directed Energy Weapon strokes in a program that does not discriminate life and of whom any can be used for non-consensual testing.

The U.S. Military Intelligence complex, in a unified effort today, along with civilian agencies, is using their electronic warfare technology to experiment upon, torture and even murder any American citizens whom even they take a disliking to, or becomes a threat in some way, or today due to technology also operating at the local police levels, police brutality has been taken to a whole new covert level in the hands of, in some cases racist law enforcement.

This is occurring while the three branches of the U.S. Federal Government completely ignore and condone these crimes, through legalization of what is happening by specific laws. That is especially true if someone makes bogus allegations in the ongoing search to test subjects.

Excerpt, "The Mother of All Black Ops", James F. Marino, excerpt, April 5, 2012

"Consider the damage done to us as a people by the US Congress, CIA, FBI, DOD, NSA and DHS alone; the biological, chemical and electronic warfare that they have subjected so many of us to; the illegal and humiliating NSA satellite spying of us within the privacy of our own domiciles; the inhumanity of their mind control torture; the complete manipulation of the US economy through the Illuminati controlled Federal Reserve Bank and IRS, and the treasonous crimes committed by them through the improperly ratified and unconstitutional Federal Reserve Act; the US Federal Government's usurpation and manipulation of the US Media through the CIA's Operation Mockingbird; the abject treason of the attacks on 9-11; and one can only arrive at the following conclusion: only a population that is either extremely ignorant of the facts, or masochistic, would ever allow for a government which is such anathema to exist within their country." -- James F. Marino - Target of MKULTRA/NSA Signals Intelligence Black Op's."

CHAPTER TWO
In His Own Words

AUTHOR'S NOTE:

The term "surveillance" is used in different ways. A literal definition of surveillance as "watching over" indicates monitoring the behavior of persons, objects, or systems. However, surveillance is not only a visual process which involves looking at people and things. Surveillance can be undertaken in a wide range of ways involving a variety of technologies. The instruments of surveillance include closed-circuit television (CCTV), the interception of telecommunications ("wiretapping"), covert activities by human agents, heat-seeking and other sensing devices, body scans, scalar waves, biometrics, technology for tracking movement, and many others.

According to Bell et al., "Sometimes, improbable reports are erroneously assumed to be symptoms of mental illness," due to a "failure or inability to verify whether the events have actually taken place, no matter how improbable intuitively they might appear to the busy clinician." They note that typical examples of such situations may include:

1. Pursuit by practitioners of organized crime

2. Surveillance by law enforcement officers

3. Infidelity by a spouse

Quoting psychotherapist Joseph Berke, the authors note that "even paranoids have enemies." Any patient, they explain, can be misdiagnosed by clinicians, especially ones with a history of paranoid delusions.

The fact is May did not have a history of paranoid delusions nor do thousands of targeted individuals being targeted by this advanced weapon system. This is the ace in the hole for its ongoing use continues through well-honed deceptive media production for the public.

Below in May's own words, is perhaps the most revealing letter of all. It is May's attempt to explain how he was pushed over the edge. The names have been deleted to protect the innocent:

MY EXPERIENCES AS A TARGETED INDIVIDUAL

My deepest regret is that I did not make a more diligent effort of documenting my experiences as a targeted individual along the way; however, this document is my feeble attempt at recounting my experiences thus far. First off, to anyone that may read this document, take a brief moment to pray for my soul. What I am about to do I have deep regret for; however, I feel that my options are extremely limited. Because I am a targeted individual, everything has been taken away from me. I have literally been robbed of life through psychological, financial, and emotional hardship.

I first realized that I was a target while I was working as an Assistant District Attorney for the Third Judicial District Attorney's Office in Dona County, New Mexico. I tried going to the police about my situation. On or about September 7, 2014, I went to the Las Cruces Police Department to make a report of my experiences. The interior of the LCPD office was locked, but there was a phone in the vestibule, which dispatch answered. I informed dispatch that I wanted to make a police report, and they sent an officer out to my location. I believe the officer's name was Name Deleted (Caucasian male that appeared to be in his 20s). The officer took notes of my experience; however, no police report was ever made. Nevertheless, there should be a record of my phone call with Dispatch. That morning, I got invited to the gun range by my girlfriend's friend's husband, Name and Email Deleted. When I met up with him later that morning at Starbucks, I informed him and another guy that was with him of the police report that I had just made. If you need to get in contact with Name Deleted, you can contact my ex-girlfriend, Name and Phone Number Deleted. I don't remember the other guy's name, but he is an engineer from Michigan (Caucasian male in his late 40s or early 50s), and his wife's name is Name Deleted. (Kenyan woman who appears to be in her early 40s, with a bald haircut).

My apartment was broken into, and my phone was tampered with. I started being followed by various individuals in unmarked cars. And on one occasion, I was followed by an individual in an LCPD sport utility vehicle cruiser. Through electronic harassment, these individuals convinced me that I was guilty of a crime. As a result, I attempted to turn myself in at the Dona Ana County Detention Center on three separate occasions. I was literally escorted to the jail by about ten cars; however, no one went inside the jail with me. Each time I attempted to turn myself in, the cars waited in the parking lot (watching me go inside the jail). But the jail informed me that they had no paperwork for me. Then, each time I left the jail after being turned away, the cars that escorted me into the jail were gone.

I continued to be followed. I informed my girlfriend, Name Deleted, who resides in Las Cruces Phone Number Deleted and my friend, Name Deleted, who also resides in Las Cruces Phone Number Deleted, of these instances. Because I was an over-achiever in my position, I frequently worked late. When I was in the office alone after hours, I would consistently see individuals peeking around corners at me. As a result of this harassment, I eventually resigned from my Assistant District Attorney position and traveled to Houston, Texas to get my old job back. I met with my old boss, Name Deleted, on the evening of Friday, October 10, 2014, to discuss rejoining his law firm, Name and Info Deleted. On that evening, Name Deleted offered me my old job back at a base salary of $50,000 per year plus 20% commissions on all settlements or judgment obtained on my cases.

After we had reached terms on that agreement, Name Deleted called his driver, Name Deleted [stout, Caucasian male who appeared to be in his late 40s], to take us to another place because we had been drinking and should not have been driving. Before Name Deleted arrived, Name Deleted friend Name Deleted arrived. It was unclear to me how Name Deleted could have possibly known who we were, but he approached Name Deleted and I at the bar and claimed that he was there to meet with Name Deleted. When Name Deleted arrived, we had one last drink. As we were leaving, I went to my car to retrieve my ID and wallet. I returned to find Name Deleted and Name Deleted whispering to Name Deleted. When I returned, everyone acted very unusual toward me; it was not the same jovial conversation that had preceded this occurrence. Then, we loaded up into Name Deleted SUV to go to the next location. As we were driving to the next location, Name Deleted leaned over and asked me, "If I wanted to take a bump"—suggesting that if I wanted to take a bump of cocaine Name Deleted or Name Deleted could get it for me. I vehemently assured Name Deleted that I did not want to take a bump. Nor had I ever taken a bump.

On Saturday, Name Deleted would not respond to my phone calls or text messages, which was strange because for the two days prior to that he had been very responsive. Name Deleted distance did not arise until after his private discussion with Name Deleted and Name Deleted. As a result of this strange behavior, I informed Name Deleted via text message that I may have to depose him in a lawsuit. Thereafter, Name Deleted called me to arrange another meeting. According to Name Deleted, he wanted me to discuss my re-joining the firm with his partner, Name Deleted. We arranged a meeting for that Monday, October 14, 2014. On Monday, we met at the A-loft, which is located at 5415 Westheimer Road next to Name Deleted law office. While there, we had drinks. Name Deleted and Name Deleted had two drinks, and I had one drink. Then, we went across the street to The West End Bar & Grille, which is located at 5320 Westheimer Road to have a meal. Shortly after we arrived, Name Deleted showed up and suggested that we go to another restaurant down the street. Because Name Deleted agreed to pick up the dinner tab, I agreed to go. At the restaurant, Name Deleted kept insisting that I join him outside to discuss matters with him. He again brought up the conversation about me taking a bump of cocaine, and I again insisted to him that I have never taken cocaine.

After dinner, Name Deleted dropped me off at my car, which was parked near The West End. The following day, I attempted to call Name Deleted, but my phone number was blocked. I attempted to call Name Deleted a couple of times after that, but my phone number was still blocked. While I do not have last names for Name Deleted, the driver, or Name Deleted, I am certain that whatever was said in that whisper cost me my job. This situation was all a part of an elaborate scheme to put me in financial jeopardy, thereby making me more amenable to harassment and less able to fight the issue. Upset and jobless, I returned to Las Cruces to continued harassment and being followed.

On the morning when I was leaving Houston, my tire was punctured and flattened while in the driveway at Address and Name Deleted, where I was staying. When I returned to Las Cruces, I had to take my tire to On Sale Tires to have it repaired. There should be a record of that repair with On Sale Tires. Further, after returning to Las Cruces, I was subjected to a constant noise campaign designed to induce sleep deprivation in me. My neighbors played loud music, screamed, and people frequently made noise right outside my windows. Not only did this happen at night time, but this happened during the day as well. Maintenance staff, which almost never worked outside my window previously, worked and made noise right outside my window every day.

The Targeting of Myron May

In addition, through electronic harassment, my life was constantly threatened. Fearing for my life, I decided to pack my things and move back home to Wewahitchka, FL. As I was loading my Uhaul to pack my things on October 21, 2014, my life continued to be threatened through electronic harassment. Additionally, as I was loading my Uhaul, there was a middle-aged man parked in the arroyo across the fence from my apartment complex (Quail Ridge Apartments – 251 North Roadrunner Parkway). He was driving a gray Dodge Durango 4 x 4 with a camper shell that had a Colorado License Plate, License Plate Deleted. This man got out of the driver-side of his truck, removed a blanket from the camper shell, and unsheathed a firearm from the blanket. Then, the most frightening thing happened. The man pointed this firearm at me, as I stood inside the trailer portion of the Uhaul truck. I jumped off the truck, ran towards my apartment, and called 911. There should be a record of this call with dispatch.

While officers were enroute to my location, the man wrapped the firearm back in the blanket and put it in the right side of the camper shell. Five officers responded to the scene of Quail Ridge. These officers were acting very aggressively towards me as though I was the assailant and not the complainant-victim. Unfortunately, I was only able to get four of the officers' names: Officer Name Deleted; Officer Name Deleted; Lieutenant Name Deleted; and Officer Name Deleted (who aggressively refused to give me his first name but whose badge number was Badge Number Deleted). Officers claimed that they searched the mysterious man, but I watched this alleged search with my own eyes. They never searched his person. Nor did they even go to the right side of the vehicle, where I saw this man place the gun. They merely peeked inside the driver-side of the pickup truck, which should hardly be characterized as a search.

After talking with the man, a little while longer, they came over to me and claimed that the man did not have a gun and that they had conducted a thorough search of his person and the vehicle. The continued to speak to me very aggressively and insisted that this man had driven from Colorado to Las Cruces, New Mexico to survey the land that he was on. The land that he was on is located directly behind Golden Mesa Las Cruces Independent Senior Living Community. If a survey of that property was requested, there should be documentation somewhere. I am certain that if you search, you won't find any such documentation. It does not even make sense that someone would travel all the way from Colorado to conduct such a survey, especially in light of the fact that you have to be licensed in New Mexico to conduct surveys. While it's certainly possible that this man could have been licensed in New Mexico, it is unlikely that Golden Mesa or the surface owners of the property behind Golden Mesa would not have simply hired a New Mexico-licensed surveyor. This experience, particularly the aggressive behavior of the officers, convinced me that law enforcement was very much involved in my harassment.

I am not sure what information was given to the leasing office at my apartment complex, but shortly after this incident, I received a letter on my door informing me that it was a notice to vacate my apartment. There should be a record of this notice and/or any interactions with law enforcement in the Quail Ridge leasing office. The harassment did not stop on this day with that incident. My life continued to be threatened via electronic harassment. I was literally threatened to leave Las Cruces immediately. As a result, I left several of my belongings in my apartment at Quail Ridge, which was Apartment 1407. The leasing office should likewise have information regarding me leaving my belongings on both the interior and exterior of my apartment. One of the things that I left in my apartment was my old cell phone (a red Samsung Galaxy S3). I had gathered several license plate numbers in this phone, but because I was told through the electronic harassment that I would be left alone if I left Las Cruces and left my phone in the apartment, I did. I, however, had forwarded some of this license plate information to Name Deleted via text message and email. In addition, I gathered some license plate information in the iPhone that was issued to me by the Third Judicial District Attorney's Office. I sent a Preservation of Evidence email to Name and Position Deleted personnel with the Third Judicial District

Attorney's Office, on October 21, 2014, in reference to preserving the information in that cell phone.

While traveling from Las Cruces, NM to Wewahitchka, FL, I was repeatedly tailgated and cut off by both patrol cars as well as regular cars. I was almost run off the road at one point in Kerr County, TX. Further, while I was in Kerr County, Texas, I was pulled over and officers searched my vehicle for no reason whatsoever. I did not give officers any consent to search my vehicle. Nor did they show me a warrant. They made me get out of the vehicle. They handcuffed me, and I stood in the heat with no shoes on for approximately three hours. Several officers were present, including Name Deleted and Deputy Name Deleted. There were about six more officers, but unfortunately due to the aggressiveness and hostility that the officers showed me, I was not able to get the others' information. Nevertheless, there should be Kerr County dispatch reports of the other officers that were involved. The officers held me there while a K-9 unit arrived from another county. My belongings were packed in plastic bins, and they literally opened every single plastic bin and searched it on the false claim that the K-9 had alerted to something inside the truck. The dog seemed more interested in the toy that the dog's handler was holding than anything in my Uhaul, and of course, they never found any narcotics.

I took a rest break in Houston, Texas on my way to Florida. I stopped at the residence of Name Deleted again. While there, the noise campaign continued. In fact, they recruited the Name Deleted family to participate in my harassment. Name Deleted (pronounced Name Deleted) is typically a quiet and reserved person, but he was continuously speaking very loudly and screaming throughout the evening. I was sleeping on the couch in the living room. While he was in his bedroom downstairs, he had his bedroom television blaring. I believe that they were recruited to participate in my harassment based on a threat of Name Deleted going to jail for possession of marijuana. At about 3:00 am that morning, while I was sleeping, Name Deleted came out of the room screaming about her daughter, Name Deleted car being repossessed. While it's certainly plausible that Name Deleted car, a red SUV, could have been repossessed, Name Deleted went on and on about the repossession—slamming doors, dropping things, and yelling at the top of her lungs. Eventually, realizing that I was not going to get any sleep in that house, I got up and continued my journey to Florida.

Of course, the electronic and other harassment continued for the duration of my trip. I was continuously cut off and tailgated in traffic the entire way. I eventually made it to Florida. When I got to Wewahitchka, Florida (and since I have been in Florida), the harassment has continued, mainly through electronic (Voice of God – see Dr. Robert Duncan) harassment. This harassment has been very debilitating and impactful. They have recruited other individuals to participate in harassing me, including Name Deleted (older, Caucasian male that is married to Name Deleted), Name Deleted (son of Name Deleted), Name Deleted (who is my cousin and the brother of Name Deleted and Name Deleted), and Name Deleted (not sure what his last name is, but he is the brother of Name Deleted). There have been others that have been sent out to stalk me as well. Unfortunately, I don't know these other individuals' names. On one occasion, about two weeks ago, I was in Rich's IGA standing in front of a three-foot, front-facing section of Little Debbie snack cakes. Two gentlemen came into the store and walked directly over to me. One of the gentlemen stood inordinately close to me, very much in my personal space. When I turned to my right toward the direction of that man, the other man came up on my left, bumped me really hard and aggressively, and began a conversation with the man on my right. The man that bumped me did not say excuse me and stared at me as though he was challenging me to say something back. Although I don't know his name, I recognized the man on my left to be one of the Gulf County Sheriff's Office deputies.

In addition, I mailed Notice of Claim letters to the Dona Ana County Sheriff's Office, Las Cruces Police Department, New Mexico State Police, and Third Judicial District Attorney's Office on October

13, 2014. I am not sure what the status of those investigations are right now. But that event seemed to have been a catalyst for my electronic harassment being ratcheted up to the point of unbearable.

Over time, I have been collecting individuals' license plates that have been following me. As previously stated, some of these have already been transferred to Name Deleted via text message to Phone Number Deleted and email. The only person that I can personally identify that was stalking me in Las Cruces, NM is Name Deleted, who resides at Address Deleted. On November 14, 2014, I got hit with a directed energy weapon. It is difficult to explain exactly what this feels like. I was also stalked by a senior-level, law enforcement officer in the Dona Ana County Sheriff's Office. Unfortunately, I cannot remember his name. He is a Hispanic male that appears to be in his late 40s or early 50s, and I know that he has a teenage son that plays football. In one discussion that we had, he stated that he worked between two candidates for sheriff on the organization chart, Name Deleted and Name Deleted. His name is either Name Deleted or Name Deleted. I can't remember. As I previously stated, I wish that I had done a better job at keeping records of the harassment that I have endured, but this is the best that I have.

SOME OF THE LICENSE PLATES

(**NOTE:** License Plate Number Deleted).

1. older model White Trans Am
2. Chevy Suburban
3. Red Honda Sedan
4. Blue Ford Escape
5. Temporary Tag –White Santa Fe
6. Chevy (followed me into a gas station just outside of Fort Stockton, TX)

7. Mexico License Plate – White Dodge Journey (followed me into post office in San Antonio)

8. White Toyota Tundra (followed me in San Antonio)

9. Dodge Charger (followed me in San Antonio)

10. White Chevrolet Silverado (followed me in San Antonio)

11. Silver Jeep

12. Dodge Durango

NOTE: License plate number deleted by Renee Pittman M. below

13. TX – Plate Numbers Deleted Here And Below

14. TX –

15. AZ –

16. NM –

17. NM –

18. NM –

19. NM –

20. NM –

21. TN –

22. TX –Black Chevy Silverado

23. *Red Toyota Yaris

24. *Older White Honda Accord

25. *Newer White Honda Accord

26. *Silver Chevy Malibu

27. *White BMW

*I did not get a state for the license plates with an asterisk beside them in Nos. 23-27, and it has been too long since I wrote them down to remember what state they were from. But I believe that they are either New Mexico or Texas license plates.

*The lady in the "20141015_133953.jpg" file followed me into the post office as I was filing the notice of claims.

*This lady in the "20141015_140529.jpg" file that worked in the post office was approached by a man that did not come into the post office for any service. The man whispered something to her. Then, when I was being waited on, she was acting very strange. I was at the post office to send my notices of claims to the Las Cruces law enforcement entities. I was mailing the Notice of Claim that was addressed to Name Deleted "restricted" mail so that he would have to sign for it. She covered up the restricted portion on the envelope. I asked her why she covered it. After I asked her that, she looked at me very surprised and swore that it was a mistake. Then she stamped restricted on the front of the envelope. But when the postage paid label printed, she covered the restricted stamp up, so I had to ask her again about covering up the "restricted" portion. And she stamped it again. In the midst of this craziness, I decided to take her photograph.

**ADDITIONAL text messages can be retrieved via AT&T

[NOTE: Post Office images and phone numbers also deleted by author Renee Pittman M.] or from Name Deleted.

OTHER TARGETED INDIVIDUALS

1. Randy Quaid
2. Melinda Fee
3. Stephen Shellen
4. Gloria Naylor
5. Kola Boof

6. Jill Anjuli Hansen

7. Matt Barasch

8. Ted Gunderson (former Senior FBI agent/whistleblower)

9. Jiverly Wong

10. Aaron Alexis

**Although I fear that they may modify or delete information in my email and/or change the passwords, here is my email information:

- myronmay@hotmail.com
  Password Deleted

- myronmay@yahoo.com
  Password Deleted

- comm.you.nick.8@gmail.com
  Password Deleted

**My Facebook account login information is as follows:

- myronmay@yahoo.com
  Password Deleted

**I believe that I removed the lock code on my phone; however, if for some reason there is still a lock code on my phone, the lock code is XXXX.

NOTE: On the Morning of December 17, 2014, after forwarding the draft manuscript to Amazon for review, I decided that I needed to verify the emails above as authentic. This was especially true regarding the Gmail, comm.you.nick.8@gmail.com. This is, as I stated, the email in which May forwarded and held the six letters. I tried the typical free email look up and was unsuccessful in attaching Myron May's name to the Gmail account. However, when I phoned Google Gmail, customer service, this account listed under the name Commu Nicate was factually listed as belonging to Myron May, and his phone number.

However, while on the line with customer service, I was told that within the hour there was activity occurring in the account giving the impression that it was being hacked. When I got off the phone, the account was closed.

Although the operation, which factually also monitors computer, keystrokes, internet connections, etc., felt this victory, the fact is, it really does not matter. They cannot delete the names, phone numbers, of friends and associates listed in the letters which substantiate the information as authentic. Nor can they delete factual evidence reveal in May's "To Do" list of how to handle his personal belonging, how to handle his property, storage unit, information of which May would only have knowledge of.

As the news reports spilled in regarding the tragedy, a colleague of May's would reveal in the excerpt below his perception of May. Interestingly, this person among many is one of the individuals May lists as playing a role in the harassment in the excerpt below:

December 1, 2014

"Name Deleted said he admired Myron May when first he met him in 2012 as opposing counsel in an employment law case. The young big-firm lawyer had an inspiring background story:

He'd lived through poverty and family discord; graduated with a law degree from Texas Tech University Law School in 2009; and earned a lucrative career defending management in labor disputes.

"There can be someone who has mental illness down the hall from you and you not know it. There was no indication that he would be capable of doing that,"

This person is one of the people May lists as playing a role in the stalking and hopeful entrapment in his letter."

It should be also noted that the Warrantless Spy Program is part of approved Electronic Surveillance activities. This includes wiretap, GPS units, and bugs, attached to vehicles, inside homes, along with around the clock satellite real time surveillance. One of the many reasons of which a person can be tracked in this program and of which is being bogusly used, in many cases, is if allegations are made of drug activity. This is whether the person is involved or not.

Today within the Militarized Police State, this opens the targeting up to psychological electronic (psychotronic) and psychophysical technology testing legally, albeit immorally, for precrime actions.

One of the laws legalizing non-consensual testing of technology for military and law enforcement personnel according to the "Exception Clause" of U.S. Code, Title 50, Chapter 32, and Section 1520a designates testing for riot/crowd control purposes. To view others DOD regulations, Acts. Executive Orders, etc., which also are connected to these efforts, legally, and without vital ethical oversight, see the "Technology Approval" tab located at the website which list several laws: bigbrotherwatchingus.com.

Can you imagine, how crushed Myron May must have been if he factually was not using cocaine as the ongoing harassment from cops seem to believe he was and escalated? An allegation of this sort, especially for an African American is a career destroyer in and of itself. This tag is believable because the African American, and also other communities, for many years has been infiltrated with drugs which in reality are another method of mind control designed to keep people from advancing to their highest.

NOTE: If there was even the remotest possibility, and there is a good chance he was not, that Myron May was using drugs that is what Rehabilitation is for of which millions from all walks of life recognize.

Psychotronic Weapons
- Brain Manipulation From a Distance -

Renee Pittman

CHAPTER THREE
Letter to Congress

One of the last letter's May drafted is below. May hoped it would be sent on behalf of all Targeted Individuals after he was gone:

Date

The Honorable _____

Office Address

United States House of Representatives/United State Senate

City, State, Zip

Dear Representative/Senator _____:

After hearing about the story of Targeted Individual Myron May and other targeted individuals, I have accepted the challenge to do what I can to make a difference. That's why I am urging you to pass a bill or resolution to begin an investigation into the targeting of United States citizens for the types of harassment outlined in my letter below:

Although not widely known by the public, targeted individuals endure a scheme of conduct designed to drive them over the edge mentally, physically, and emotionally. This scheme is referred to among the targeted individual community as: gang stalking." Gang Stalking is a systemic form of control, which seeks to control every aspect of a Targeted Individuals life. Gang stalking takes place outside in the community. It's called gang stalking because the target is followed around and placed under surveillance.

Gang Stalking Goals

The goals of gang stalking are to (a) silence an individual, (b) drive a victim insane and possibly to the point of suicide, and (c) destroy the victim's reputation and believability—as the person will be viewed as mentally ill if they complain or report such abuse. Gang stalking may also be used to force an individual to move or leave an area.

Motivations for Gang Stalking

The motivations for gang stalking vary widely. Three of the most common are as follows:

- Being a corporate or government whistle blower (particularly if you are exposing conduct that is unbecoming of law enforcement);
- Revenge for getting away with a criminal offense;
- Knowing too much about gang stalking itself—because you become a threat to exposure

Who Are The Stalkers?

For the most part, stalkers are everyday citizens that are usually acting under some type of guise (e.g., that they are doing something positive for the community; a false accusation that the targeted individual is under investigation for a crime). Other stalkers, however, are simply paid to harass a targeted individual. Sadly, many of these gang stalkers are involved in law enforcement.

Examples of Gang Stalking Harassment

Here are some of the stalking activities that targeted individuals have to endure: slashed tires, threatening phone calls, hangup calls, verbal assaults by strangers, property damage, death threats, peeping toms, being followed on foot and by vehicle, and character assassination among family, friends, neighbors, and coworkers.

Tactics Used by Stalkers

Anchoring

A technique employed by stalkers to persuade the targeted individual that a particular group is responsible for the abuse—neighbors, racial-ethnic group, etc. The goal is to trick the targeted individual into thinking an innocent bystander is the culprit of their harassment and compel the targeted individual into verbally or physically assaulting that person.

Brighting

Brighting involves the practice of repetitive flashing of a car's high beam headlights at a targeted individual. The targeted individual is flashed an inordinate amount of times from either a tailgating, passing, or oncoming vehicle. Brighting also occurs when bright lights are flashed into a targeted individual's home windows.

Electronic Harassment

Electronic harassment is the use of technological devices to spy on or cause harm to targeted individuals (e.g., exposure to high electromagnetic fields, microwave radiation, etc.). A frequent form of harassment involves beaming a low frequency tone into a targeted individual's area, which over time causes sleep deprivation, agitation, and stress. A great deal of this research is highlighted in a book written by the pioneer of electronic harassment, Dr. Robert Duncan, which is entitled "Project Soul Catcher: Secrets of Cyber and Cybernetic Warfare Revealed." Other prominent individuals in the field include Dr. Nick Begich, Dr. John Hall, and Dr. Terry Andersen. In addition to electronic harassment, stalkers utilize nanofibers to track targeted individuals.

Ghosting

Ghosting refers to the practice of rearranging or moving a targeted individual's personal belongings to make the targeted individual question his/her sanity (e.g., moving home furniture, lawn decorations, desk decorations, etc.).

Mimicry

Mimicry is a specialized form of harassment in which the stalkers imitate every movement made by the victim.

Mobbing

Mobbing is a term that describes intense group bullying. Several stalkers descend upon a targets area in the same time period. In these instances, gang stalkers are not as discrete because they want to make their presence known to the targeted individual.

Noise Campaign

A noise campaign is an orchestrated effort to produce stress in a targeted individual by prolonged exposure to noise (e.g., neighbors playing loud music, cars passing at rapid paces, vehicle horns blowing, technological devices to induce dogs to bark, etc.).

Sensitization

Sensitization is a psychological term referring to the forced association between a stimuli and a corresponding reaction. For example, if a targeted individual is frequently harassed by people wearing blue baseball caps or sunglasses, over time, the targeted individual will believe that anyone wearing a blue baseball cap or sunglasses is a stalker coming to harass them.

Gas-Lighting

Gas lighting refers to presentation of false information for the purpose of making victims doubt their own memory, perception, and sanity.

Hacking

Hacking generally refers to the downloading of viruses and malware as well as gaining remote access to a targeted individual's electronic devices. After gaining such access, stalkers are able to delete files, modify files, and contemporaneously interfere with a target while he or she is working.

It is my hope that you will accept the challenge as I have. Again, I strongly urge you to consider passing a bill or resolution to conduct an investigation into the harassment outlined in my letter above. Thank you for your consideration.

Sincerely,

Name

Title

Address

City, State, Zip

Phone Number

After posting of this letter to various social networks, one person took offense to May's description of "Motivation for Gang Stalking" in this hopeful letter to government officials as highlighted below:

Motivations for Gang Stalking:

The motivations for gang stalking vary widely. Three of the most common are as follows:

- Being a corporate or government whistle blower (particularly if you are exposing conduct that is unbecoming of law enforcement);

- Revenge for getting away with a criminal offense;

- Knowing too much about gang stalking itself—because you become a threat to exposure

The fact is community groups are factually being motivated after they are falsely told the target is a criminal, on drugs, or mentally ill! No one wants either in the community.

To anyone who sends this letter out, you are welcome to point this fact out if you so desire. However, the fact is, people are being denied, moral, Constitutional, Civil and basic Human Rights. Another targeted individual made it clear saying "I don't like psychopaths, but I don't believe they should be horrifically tortured either before or without moral judicial court proceedings. This factually leaves human lives in the hands of ego driven psychopaths who believe they are right all the time even when wrong.

I however, believe that Myron May made it clear in the next paragraph that organized stalking, he had learned to be a law enforcement mobilized community effort of which it factually is. Many recognize this as why when calling for police help, the target can be harassed, ridiculed, intimidated, by police and with no support or police reports filed for documentation:

Who Are The Stalkers?

For the most part, stalkers are everyday citizens that are usually acting under some type of guise (e.g., that they are doing something positive for the community; a false accusation that the targeted individual is under investigation for a crime). Other stalkers, however, are simply paid to harass a targeted individual. Sadly, many of these gang stalkers are involved in law enforcement.

Another factor is the Department of Defense intentionally passing this military technology to the Department of Justice thereby allowing combined military/law enforcement use after 9/11. In fact, 9/11 was officially used as the foundation to evolved this program's legalization overall in the hunt for terrorist. However, it appears that the technology has been factually turned on US citizens and those privy to its use have become technological terrorist within the USA.

There are also many who have been targeted of which many are recognizing from birth. This program did not just appear overnight but has been an ongoing mind control program documented to be years in the making. Some, who did not know they were targeted prior to energy weapon widespread use, have been set up, or nudged into crimes by those sitting in these advanced operations in real time toying with and destroying human lives. Population and group influencing and manipulation that occur through biometric algorithms set for specific communities.

In fact, labelling anyone as a criminal or a possible criminal is factually the intent of the Department of Homeland Security's Malintent software. It scans people for possible intent to commit a crime and a person can be targeted for a passing thought. With the Thought Police now at the helm, they are trained that if someone thinks "I hate government" or police, they can be targeted, labelled a possible domestic terrorist, blacklisted, technologically harassed, and attacked by Directed Energy Weapons, etc.

I believe that May was quite accurate detailing various reasons in his limited time in this program and because of this; he did not grasp the full spectrum of various types of ongoing research using various people for a variety of reasons.

CHAPTER FOUR
"Strange Night of November 19th"

Whenever I look at this title, of one of the letters in the Gmail, I ask what exactly did May mean when he labelled it "Strange Night of November 19, 2014?" Is there some clue to his mindset prior to the shooting rampage? I could speculate and say that he was possibly feeling some type of psychological electronic effect as the time drew nearer and nearer but that does not make it right. If this were the case then it could make what happened later make sense?

NOVEMBER 19, 2014

- I sent ten packages priority mail, certified mail, return receipt requested to various recipients. All recipients are indicated in the file titled "Letter to Renee and Others"

- Nine of the ten return receipts are going to Renee Pittman Mitchell ADDRESS DELETED; the other return receipt—for the package going to Renee Pittman Mitchell herself—is going to Derrick Robinson.

- All ten of the certified mail tracking numbers is listed here. Mysteriously, the tracking number for Renee Pittman Mitchell is not listed. All of the tracking numbers are in these files:

> Screenshot_2014-11-19-17-56-01.png

> Screenshot_2014-11-19-19-02-55.png

> Screenshot_2014-11-19-19-03-05.png

- In the following three photos, you will see the certified receipt for the letter addressed to Renee Mitchell (first), the actual package addressed to RM (second), and proof of the $11.75 postage paid for all 10 packages in (third – fifth)

Earlier I mentioned that I thought that May was an imposter. Below by example is the reason that I did not respond to Myron May's after around 6:00 CST on November 19, 2014.

As shown below, my decision was made after May sent images of the postal receipts for the certified mailings. It is my understanding that typically the post office stamps, as confirmation the green and white receipt. This is shown by the random image I used as an example on the next page. As noted, the receipt is date stamped and also the actual cost for postage for the certified mailing is documented also on the receipt.

None of the receipts which May sent images of beforehand for substantiation, had what I thought was typical, date, time, or price information stamped on it for what I believe should authenticate.

Certified Mail Receipt from USPS website clarification:

If you mail a letter USPS Certified Mail the Post Office provides the mailer with (2) mail receipts. The first certified mail receipt is many times described as 'proof-of-acceptance' this is a receipt from the Post Office that they accepted your letter on a specific date and time to prove your letter was mailed. The second is referred to as 'proof of delivery'

When you mail a Certified Mail letter you may need proof that you mailed the letter timely as well as evidence of delivery (to whom the mail was delivered and date of delivery), along with information about the recipient's actual delivery address. Certified Mail forms are available free from the local Post Office. If you are mailing more than 3 items USPS suggests using a 'firm sheet' or manifest (PS-Form 3877). This special certified mail receipt is printed using computer software and was designed to help large volume mailers that send lots of USPS Certified Mail. The software insures the USPS Certified Mail article number is correctly listed with the accurate delivery address for each recipient with the unique article number for each piece of Certified Mail. Additionally, all USPS special service fees including the certified rates and costs, first class postage and additional endorsements are correct according to USPS standards.

When you take your letters or articles to the Post Office window clerk or Business Mail Entry Unit (BMEU) make sure they round stamp (date stamp) and sign the firm sheet or PS-3877 manifest. This becomes your receipt and legal proof of mailing and acceptance of your letter into the USPS mail stream.

No other paper receipt will be provided by USPS unless you are using the Electronic Return Receipt that provides electronic tracking and delivery confirmation service.

N20210523085842509

Renee Pittman

Example of an officially stamped USPS Certified Mail Receipt

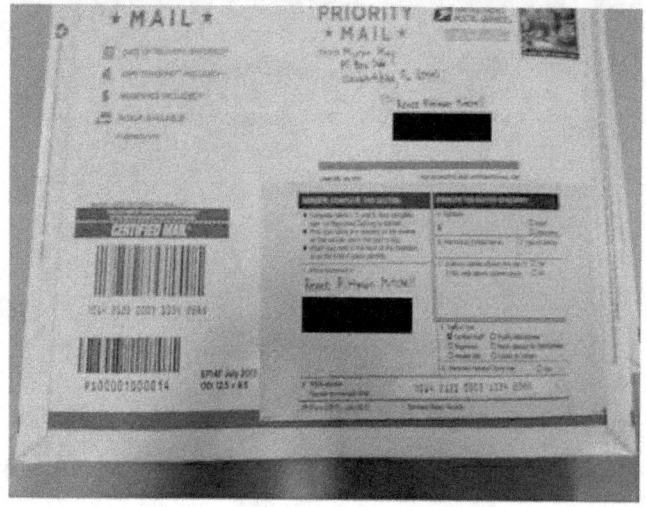

I also felt that the images below which appeared to be receipts for stamps did not prove certified mailing of packages.

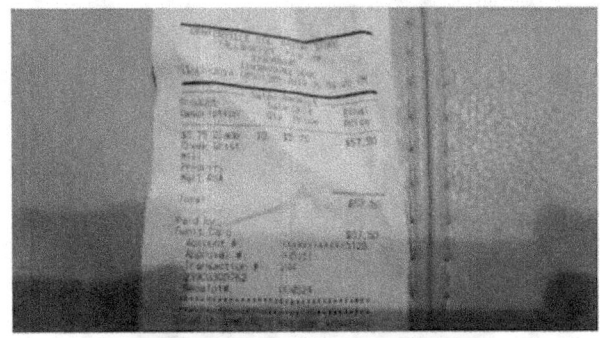

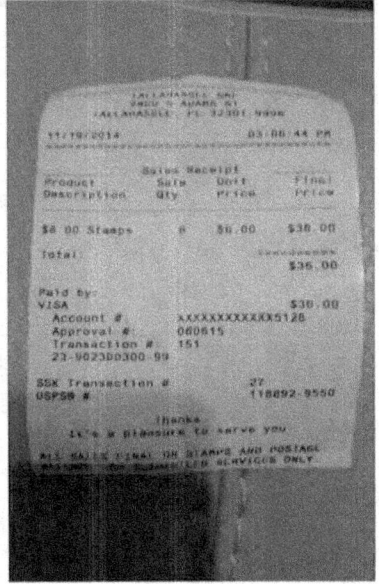

Below are images of text message from Myron May which he hoped would further substantiate certified mailings. Following the text, images he explains what they mean.

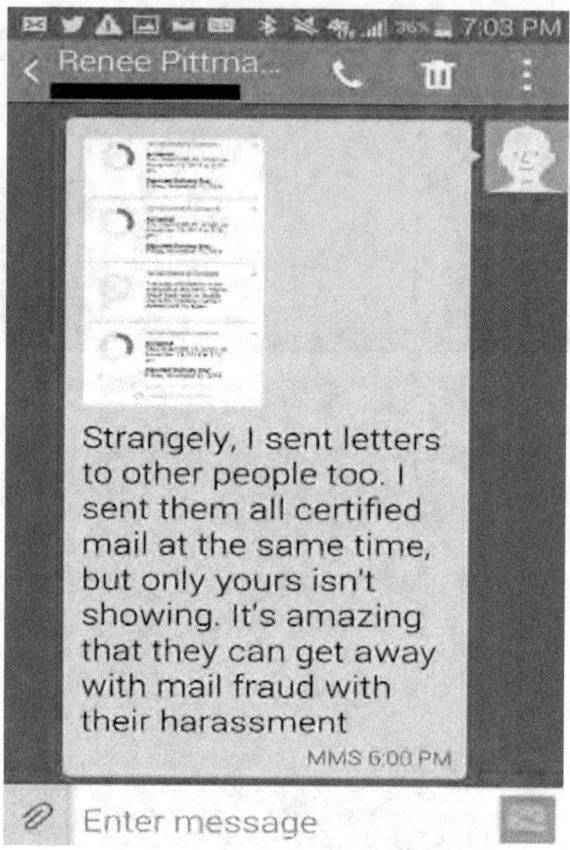

In the above screen shot, I am sending a text message to RM along with an image of her tracking number attached showing that her certified letter is the only one that did not go out. I anticipate that they are trying to intercept the other packages.

I am unsure if this is the PS Form 3877 because it is illegible to me and is unfamiliar to me as something I use when mailing certified documents.

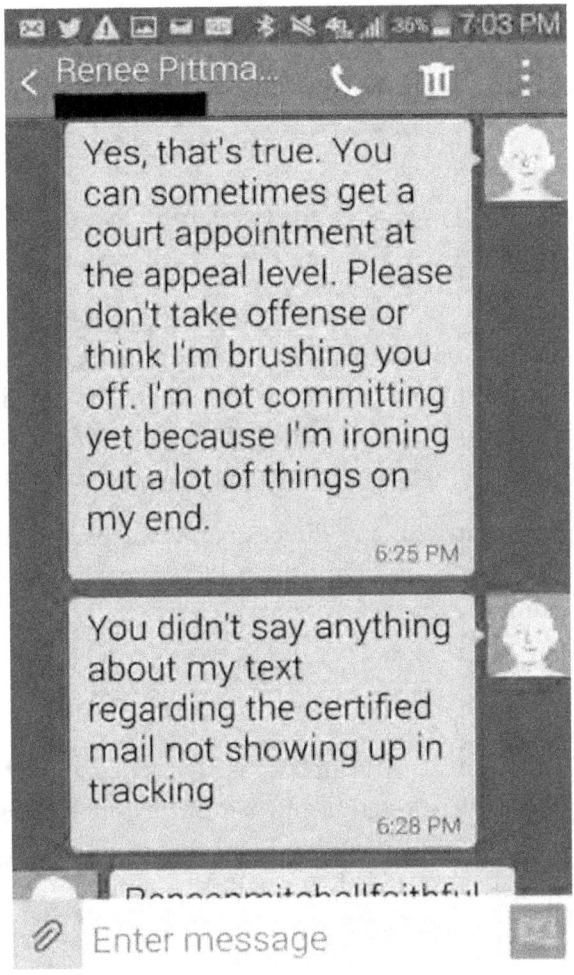

In the above text message, I am asking RM about the message that I sent her b/c she is typically very responsive and she was talking to me about other things. It became apparent that my message had been intercepted. I called RM, and she confirmed that she did not receive the message regarding the package that was being sent to her disappearing. Immediately after that phone call, I sent them via email to her, as evidenced below:

This is the last email from Myron May sent 9:16 p.m. CST of which I did not respond, because again, I did not see any typical post office confirmation stamps on the green and while postal receipt as stated previously.

Email to my yahoo account:

On Wednesday, November 19, 2014 8:16 PM, may Myron <myronmay@yahoo.com> wrote:

There are 3 important videos that I made and mailed. Make sure the stalkers don't steal them. Renee, you should get the tracking number updates from the post office. I added your email to the tracking updates. I also sent you photos on Facebook of all that stuff.

I've been getting hit with the direct energy weapon in my chest all evening. It hurts really bad right now. Please make sure you get the items I'm sending.

End of email

I had to call RM again at 7:21 pm because I was being hit with directed energy weapons and felt significant pain in my chest. This pain was continuous and unlike any pain that I had ever experienced previously. The pain subsided immediately after Rose Campbell came home at approximately 8:00 pm CST.

NOTE: 6:21 p.m., PST in California

MYRON MAY TIMELINE

- 2001 - Graduates from Wewahitchka High School
- 2005 - Graduates cum laude from Florida State University; economics major
- 2009 - Graduates from Texas Tech School of law

- September 1, 2014 –Tells two friends via emails he was considering starting his own law firm.

- October 5, 2014 - Email to two colleagues

Subject: Striking Out On My Own

FIRST - I'm about to strike out on my own and hang my shingle. If I can help you with some overflow work or something, please let me know.

Sent from Yahoo Mail on Android

SECOND - I am about to strike out on my own and hang a shingle. I'm going to stay here in the Las Cruces / El Paso area. If I can help you with some overflow work, let me know.

Sent from Yahoo Mail on Android

October 7, 2014 – Ex-girlfriend, an MD tells New Mexico police May is showing symptoms of "severe mental disorder," that he quit job with the district attorney's office and felt he was being spied on.

October 12, 2014 – Myron May received an alert from Microsoft related to "Unusual Activity" on his Hotmail email account:

October 13, 2014 - Myron may details escalated stalking before filing a Tort Claim against several officials in New Mexico to a friend and how, as a result it appears his stalking intensifying:

Fw: Notice of Claims Against Las Cruces Entities People

To: Names Deleted

I also need you to save this email and remember these names:

Name Deleted (Driver)

Name Deleted (Driver's Friend)

I'm not sure about this, but I want to save it anyway. This morning, a Jeep came through the neighborhood with the license plate License Plate Deleted. The Jeep strangely came speeding around the corner and then stopped near the end of the street where it curves. There were two occupants in the vehicle, neither of which got out of the vehicle. The driver was a Caucasian female, and the passenger was an African American female. I drove toward them, and they both suddenly turned their heads away from me. When I drove back by them, the driver (Caucasian female) ducked her head in the car.

October 13, 2014 – Email giving a friend family contact information.

On Monday, October 13, 2014 4:06 PM, may myron <myronmay@yahoo.com> wrote:

Also, I want all of you to have my mom's contact information:

Cell: Phone Number Deleted

Home: Phone Number Deleted

Name Deleted Number (in case she doesn't answer): Phone Number Deleted

October 13, 2014 – Myron May detailing he is going to Post Office to mail "Notice of Claim" to begin a Tort Claim.

On Monday, October 13, 2014 4:01 PM, may myron <myronmay@yahoo.com> wrote:

Name Deleted:

I am about to go to the post office to send these letters. If I am stopped on the way, make sure that they get sent.

Thank you,

Myron

Below is the "Notice of Claim" against Las Cruces Entities. Who made false allegations of Myron May as a drug user. He sent the notice to five officials in Las Cruces, New Mexico:

Myron May

251 N. Roadrunner Pkwy., Apt. 1407

Las Cruces, New Mexico 88011

October 13, 2014

VIA US CERTIFIED MAIL, RETURN RECEIPT REQUESTED

TO: Name, Organization Address, Etc... Deleted

Re: Notice of Claim Pursuant to New Mexico Tort Claims Act Claimant: Myron May

Incident Date: 10/10/2014; July 2014 – October 2014

To Whom It May Concern:

The purpose of this letter is to give notice pursuant to the New Mexico Tort Claims Act of claims against Organization Deleted. NM Tort Claims Act, NMSA 1978, § 41-4-16(A). Myron May ("May") anticipates filing claims for injury and damages, including, but not limited to, for intentional torts of tortious interference with contract, defamation, and violations of section 1983 of the Civil Rights Act.

These claims arise from a deprivation of May's right under New Mexico, Texas, and federal law, which will include assertions that Organization Deleted employees and/or affiliates, along with local law enforcement agencies, intentionally interfered with a contract for services between May and Name Deleted. Specifically, May contends that Organization Name Deleted employees and/or affiliates, along with local law enforcement agencies, made assertions to Name Deleted that were false or with reckless disregard for their falsity, which were intended to prohibit, cancel, or impede the contractual relationship between May and Name Deleted. In addition, May contends that Organization Deleted employees and/or affiliates, along with local law enforcement agencies, orchestrated a program of surveillance in violation of the Fourth Amendment to the United States Constitution and Article II, Section 9 of the New Mexico Constitution in or around the dates of July 2014 through October 2014.

May urges the Organization Deleted to conduct an internal investigation into these matters. If the Organization Deleted is interested in resolving the claims raised herein, the Organization Deleted may contact May via telephone at Phone Number Deleted or via email at myronmay@hotmail.com. If May does not receive a response from the Organization Deleted within a reasonable time, May will pursue additional remedies at law and/or in equity.

Respectfully,

Myron May

October 20, 2014 – Myron began to do research related to electromagnetic energy focusing on Radio Frequency Microwave Interaction with Biological Tissue and located the two websites below:

Fundamentals of Electromagnetic Engineering

Fundamentals of Electromagnetics' with MATLAB 2nd Edition

October 21, 2014 – After resignation from DA's office, May sent the email below to an associate entitled "Copy and Paste of Email for Preservation of Evidence." The below email was also in the recovered Gmail.

It appears the email is related to images in the law firm's cellular phone he turned in at resignation. It appears that Name Deleted does not know what he is talking about related to the license plates and asks if the plates are related to a case:

October 21, 2014.

Copy and Paste of Email for Preservation of Evidence

Ok do you recall what case the plates belong to so I can inform the secretary of this…

From: may myron [mailto:myronmay@yahoo.com]

Sent: Tuesday, October 21, 2014 7:37 AM

To: John

Cc: (Names deleted by author)

Subject: Preservation of Evidence Request

Name Deleted:

I left several license plates under "Notes" in the iPhone that was assigned to me by the office. Please preserve this information as evidence of litigation. The cell number associated with that iPhone is Phone Number Deleted.

If this information has been destroyed, deleted, stored elsewhere, or is otherwise not recoverable, please provide date, time, and reason why such destruction, deletion, or removal took place, along with any supporting documentation.

Thank you for your cooperation.

Regards,

Myron May

October 28, 2014 – Myron May below was recommended concoction he took hoping to detox his body of nanofibers / technology recommended by someone.

Today at 11:01 AM

Subject: Vodka Wash, Zappers, and Magnets to Combat Nano-Fibers

To: myronmay@yahoo.com, myronmay@hotmail.com

Good post there, Name Deleted. Yep….that's what I got.

Hello Everyone. I just got up from my nap after doing a vodka wash from the inside…and oh what a headache. I'm not much on the hard stuff. I'll do my outside wash later. Made me a zapper yesterday…of sorts. I used two 9-volt batterys instead of the one. It's a zapper all right. Did some other experimental stuff yesterday, as I got to thinking about protons/neutrons and electricity in general. I rigged me up a hat w/magnets spaced about in the inner lining, and in between those I have #4 copper wire (I wanted to go w/a more slender gauge but did not have any on hand…I gotta get some) going into the apex. I also rigged up some shoes w/magnets, as I am setting on a magnet w/2-hundred lb weight capacity. Later on I'm going to apply the juice to the hat. Too early to comment.

Name Deleted make a paste of at least 3X filtered vodka with baking soda, keep it in a jar. Get wet in the shower, turn off the water and wash with the vodka baking soda paste, head to toe, using a bath brush helps..rinse.

The Targeting of Myron May

Diatomaceous Earth for Weight Loss, Dry Skin, Joint Pain, Detox

Myron May

October 28, 2014 – Letter to father figure in May's life

To: Email Address Deleted

Please call me as soon as you separate from the military. I'm suffering here Pops.

October 28, 2014 – Email by Myron May expressing interest in a Targeted Individual lawsuit to unknown party.

To: Email Address Deleted

I would like to join in the lawsuit against the FBI and FCC. I have gathered some information myself that could be beneficial in processing the claims. I am a lawyer as well; therefore, if I can be of any assistance, feel free to let me know. All of my electronic devices have been compromised and manipulated, so here is my contact information:

Phone: Phone Number Deleted

myronmay@yahoo.com

myronmay@hotmail.com

comm.you.nick.8@gmail.com

October 28, 2014 – Excerpt of Myron May's email to himself, related a targeted individuals. This appears to be after doing case study, research and comparisons. A partial of the actual lawsuit he located is below.

Subject: Possible Lawsuit in DC District (D.D.C.) ~ Federal Court Actions

To: myronmay@yahoo.com, myronmay@hotmail.com,

comm.you.nick.8@gmail.com

IN THE UNITED STATES DISTRICT COURTS FOR THE DISTRICT OF COLUMBIA

NOTE: Example

Name Deleted Plaintiff V.

FEDERAL BUREAU OF INVESTIGATIONS, et al. 935 Pennsylvania Avenue Northwest Washington, DC 20535-0001;

CENTRAL INTELLIGENCE AGENCY Office of Public Affairs Washington, D.C. 20505;

NATIONAL SECURITY AGENCY 9800 Savage Road, Suite 6711 Fort Meade, MD 20755-6711;

DEPARTMENT OF JUSTICE 600 E Street, N.W. Washington, D.C. 20530;

DEPARTMENT OF DEFENSE 1400 Defense Pentagon Washington, DC 20301-1400;

DEPARTMENT OF ENERGY 1000 Independence Ave Washington, DC 20585;

UNITED STATES OF AMERICA Defendants

COMPLAINT

1. Plaintiff, Name Deleted hereby brings this action for injunction relief and damages based on personal knowledge and belief, as a victim and expert witness, to the information provided, as to all other matters, as to which allegations Plaintiff, without doubt or delusion, will provide proof, unrefuted evidence, overwhelming evidentiary support, witnesses, substantial facts, research, and investigation that exists as follows:

PRELIMINARY STATEMENT

2. This case is about the wiretapping, civil rights violations, privacy, communications technology, and crimes of humanity, in use by the Intelligence Community. With the advanced technology unknown to most, of shadow network of surveillance and spying under DARPA projects, defendants are, transmitting, intercepting, tampering, and blocking the content of a significant portion of the Plaintiff's phone calls, emails, instant messaging, text messaging, www, Skype, electronic and wireless communications, and other communications, both internationally and domestic, including Plaintiff's family and practically every American, for the past 4 years, beginning on or about January 2007.

3. Plaintiff's records will show communications are intercepted, manipulated, tampered with, stored, (data mining), harassed, and impeded. Plaintiff will provide numerous emails hijacked, fabricated, UNREAD, manipulated, deleted, including facsimile, and internet manipulations, since 2007. Plaintiff's numerous impeded electronic communications with all her service providers were debilitating to her work and she had no resolve available from providers who were not able to correct or detect the situations. This includes manufactured emails, phone calls, and government employee's unwitting involvement. In 2008, Plaintiff was also accused legally of false communications during her service with Qwest, by using wiretapped calls and disconnects to create false

records with her utilities company, including defamation to her character. (Name Deleted, Military wife; whistle blow case on tampered communication CHAOS and automobile impediments not to testify).

4. Plaintiff is not delusional and has provided her records to DOJ in early 2009, with documents from many other credible sources and whistle blowers. Plaintiff requests that the Russell Tice case and Articles below § 35 is read with disclosures being made as to her allegations, Injunction relief, various separate counts, and tort damages throughout. Plaintiff has demonstrated courage, risked her reputation, risked her life, and family, and has been threatened, prior to receiving an anonymous email of the NSA article (below) in February 2009. Plaintiff tried to blow the whistle to the DA late 2007, and was not aware what NSA was or government corruption, but knew she had to report what was going on. As quoted by Mr. Tice and others; "Statement by NSA, "THE TIP OF THE ICEBERG and WIRETAPPING SURVEILLANCE LOOK LIKE SMALL POTATOES". Most recently, Plaintiff was able to meet with the FBI in early 2010, and provided some information for investigation by various FBI analysts.

5. The President and other executive officials have described some activities of surveillance and spying which are conducted outside the procedures of the Foreign Intelligence Surveillance Act ("FISA") and without authorization by the Foreign Intelligence Court, Communities, Committees, Senate, or Congress.

6. As with The Attorney General and the Director of National Intelligence having since publicly admitting that the TSP was only one particular aspect of the surveillance activities authorized by the Programs, and are being abused. (Jewel v United States, United States v Yahoo)

7. In addition to eavesdropping on or reading specific communications, Defendants have intercepted the communications contents across the United States and overseas,

and targeted the Plaintiff with intense sabotage and impediments, Electronic Stalking, Slander, including Internet Communications, and Google's Access Portals. Plaintiff wrote to Google to let them know what was going on behind the scenes with all the tampering including remote viewing, password access, and covert data-mining in SG3 dating back prior to 911. (EFF v Google and Street View) DOD's ELF transmitters were already in full scale by 1981 in Australia and Africa, followed by additional covert superior Projects as Combat Zones also under DARPA formerly ARPA)

ALLEGATIONS AND FACTUAL ALLEGATIONS RELATED TO ALL COUNTS

8. With suspicious FBI allegations and $122 Million Virtual Case File Project gone wrong in 2001, prior Board Members and Government Officials', the new Parent Company SAIC's (C I A Spacestar) servers and the ("Occult Connections"). Chief of Staff Jack E Thomas Air Force Intelligence, 1997 ties with DOD, NSA, CIA, and former Executives, Directors, Secretaries, Army Generals, all had knowledge of REMOTE VIEWING experiments spy biz, SPAWAR at the Naval Electronics System Command in San Diego and Los Alamos National Labs for Medical Oversight to American Intelligence Agencies (1996). A surge of Federal Spending raised Multi-Billion-dollar Defense concerns with SAIC and Titan (moon) in 2004 (San Diego) and the mid 90's on technology projects that REQUIRE HUMAN EXPERIMENTING as well. These BLACK BUDGETS are corruption hiding behind SSP.

9. Joseph McMoneagle, eventually revealed as such, but for the purposes of the Army's psychic intelligence unit, he was simply Remote Viewer No. 1." In his Memoirs of a Psychic Spy is a look at the most remarkable exploits in a most astonishing career of military service. Remote Viewer 001 in Army's Stargate Project reports this as still classified. The top-secret intelligence gathering initiative launched at height of Cold War- David Morehouse 1998.

This project was first used to retrieve intentions in a criminal or terrorist's mind. (The Lucid View, Investigations in Occultism, 2004 and included in MK ULTRA is Project "VOICE of GOD") This is used for unexplained phenomenon and to manipulate religion and/or to compel a crime, command or order. It is undeniable that government agencies have tested citizens without permission as The Manchurian Candidates in the United States and the IRS. Cognitive Sciences Laboratories CIA research 24 years ago when Tom Beardon warned about these weapons and activities as serious dangers.

10. The core component of these Programs is Defendants' nationwide network of sophisticated communications surveillance and spy devices installed overtly. NSA transmissions and NASA Remote Viewing have been in place and are being connected to the key facilities of telecommunications with advanced technologies like that of HAARP stations and Satellites for many years prior and covertly. As Quantum Physics (sound and/or radio WAVES) and subliminal are being used for Spying and breaching contracts with other Countries and Embassies, HAARP is used for Earth and human manipulations. (Jerry Smith, Jim Keith; Politician and Dr. Nick Begich, Alaska; "HAARP; THE ULTIMATE CONSPIRACY" 1998) The "Method of Inducing Mental Emotional and Physical States in Human Beings" was patented for the CIA in Virginia in 1993.

11. These projects of CointelPro, Psyops experiments for drones, and Surveillance/spying have continued covertly to develop weapons that were first patented by Russia and China. Since the 1950's, first revealed to the public were the Alien UFO Projects, and the subliminal cravings of coca cola, movies and popcorn. Shortly thereafter as technology advanced, mass control continued to this day nefariously against all Americans, called NSA transmissions today. Since 1974 under the Pentagon, DOD funded the ELF radio broadcasting in conjunction with hypnotic control and prior to that Doctors Bill Nelson and Tom Beardon had their own

patents at NASA on Medicinal Energy and Biofeedback, already in use in other Countries with the same remote scalar waves and frequencies. (Apollo and Whistleblowers Second craft 11, shadowing Apollo 13 and no witnesses returned after the Shuttle)

12. In early 2008, after 25 years of being an accountant, and raised in the Mediterranean, Plaintiff purchased the Non-FDA device from a NASA physicist at $21,000.00 with Scalar waves, (Russian Woodpecker frequencies), and Military Patents. Plaintiff became a certified practitioner to protect herself and attempt to deprogram. Plaintiff was intercepted by this HAARP and NSA's SIG-INT Satellites, thus became a certified hypnotist trained in NLP (Neuro-Linguistic Programming; CIA taught techniques, (Bandler and Grinder 1976) and (Project Paperclip; Dr. Strughold Space Medicine 1963)

13. In her travels, Plaintiff witnessed these transmissions used in Israel, Australia, England, European Countries, and was further impeded when attempting to whistle blow the nefarious uses and crimes of humanity, to the embassies. Plaintiff will show documents or connections to MAFIOSO, KGB, and AL Qaeda, within the CIA, as with Donald Rumsfeld and his banking ties. World Banks (1910) JP Morgan, Black Budgets, and Nikola Tesla 1940)

14. Some aspects of the Programs of MKULTRA were publicly acknowledged by President Clinton with partial apologies on October 3, 1995 as past tense and not being done currently to down play their uses. The Pentagon has new buzz words…

Nov. 3, 2014 – May continued to have bankruptcy problems.

Capital One secures judgment against May for $1,710.60. Nov. 20, 2014

November 3, 2014 – This email is related to Myron May's first contact with Targeted Individual organization, FFCHS, Director, Derrick

Robinson and Derrick Robinson's response to Myron May after May volunteered legal services to help targeted individuals

From: Derrick Robinson

Sent: Monday, November 03, 2014 4:12 AM

To: myronmay@hotmail.com

Subject: Re: Volunteer

Dear Myron,

Thanks so much for volunteering to help with our Legal Committee. I will be in contact with you soon.

Derrick

Phone Number Deleted

November 11, 2014 – This is an email May to himself holding one of many prayers for "Spiritual Warfare." May initially copied them on October 30, 2014 and then November 11, 2014 emailed back to himself to possibly recite.

Re: Spiritual Warfare Prayer

TO: Myron May

From: "Myron May" <myronmay@hotmail.com>

Date: Thu, Oct 30, 2014 at 9:42 AM

Subject: Spiritual Warfare Prayer

Lord Jesus, You promised in Your Word that "I can seek You and find You when I search for You with all my heart!" (Jer.29:13) So Lord, I now cry out to You with a WHOLE heart, and I ask You to please hear my prayer and give me the deliverance I seek! You said, "They looked unto Him and were lightened, and their faces were not ashamed - This poor man cried and the Lord HEARD him and SAVED him out of all his troubles!" -- Psa.34:5, 6.

November 14, 2014 – Myron May second email to Derrick Robinson FFCHS after not hearing from him after the November 3, 2014 email and no response until November 14, 2014.

On Fri, Nov 14, 2014 at 10:11 AM, Myron May

<myronmay@hotmail.com wrote:

I haven't heard anything back from you. What's up?

November 14, 2014 – This is Myron May's email, someone in military who again, appears to be a father figure in his life. It details him being subjected to powerful Microwave Direct Energy Weapon attacks:

To: Name Deleted

Pops,

Today, I got hit with what felt like a microwave weapon. It lasted for only a minute or two, but it hurt really bad. It felt like my insides had a painful tingle. I started sweating and got dizzy. It was so scary. It came out of nowhere. I've never went so quickly from normal body temperature to sweating.

Sent from Yahoo Mail on Android

November 14, 2014 – Derrick Robinson's email response to Myron after May questions Robinson why he had not contacted him since the November 3, 2014 from Robinson titled "Volunteer."

To: myronmay@hotmail.com

From: derrickcrobinson@gmail.com

Hi Myron. Thanks for writing. I called and left a message at the number you gave me, but haven't heard from you.

Can you call me at your convenience?

Derrick

Phone Number Deleted

NOTE: A lot of people believe that the great evil torturing and taking life, which is unprecedented and horrendous, of which targeted individuals endure is the result of possession of evil spirits. I believe that those employed in these groups are "Human Monsters." As I was told, "This is how we earn our living."

In my case, I prefer to keep it in the realm of those we can actually fight without an Exorcism!

November 16, 2014 – Below is an extended version of Myron May's research to understand the powerful, psychologically crippling, dynamic of Organized Stalking aka Gang Stalking from a Spiritual standpoint and a link he found. This is a more detailed version of information in his Letter to Congress showing how he came to his conclusions and how he attached the experiences of stalking to his life.

To: Myron May (May sent to himself)

Subject: Gang stalking and its preternatural connections.

Evil takes many forms in this world. Its areas of influence go deeper into the workings of our lives and institutions than many are aware of. We began SIS with the goal of protecting people from the manifestations of this evil, and when asked, cleansing their environments from it. Though it may be hard for some to fathom, our world is influenced, manipulated, and managed by what some refer to as Archons (Def. a ruler).

In Ephesians 6:12 we find other titles for them: For our struggle is not against flesh and blood, but against the rulers' powers of this dark world and against the spiritual forces of evil in the heavenly realms.

Here are some other titles in Romans 8:38: For I am persuaded, that neither death, nor life, nor angel principalities powers, nor things present, nor things to come,

These beings have one primary purpose. The ruination of mankind's soul. I will cover the many means they employ to accomplish this in future articles. They feed off of human suffering. It very literally is food for them. These are very grim topics, but ones which must be addressed.

Today's entry will cover a subject which is not widely known by the public. It is a tactic which is used to drive people over the edge. It is called Gang Stalking. Gang Stalking is a systemic form of control, which seeks to control every aspect of a Targeted Individuals life. Gang Stalking takes place outside in the community. It's called Gang Stalking, because the target is followed around and placed under surveillance by groups of organized Civilian Spies/Snitches 24/7. I personally dealt with this situation when I was a law enforcement officer. I was investigating a large organization of luciferians, and they employed all of the tactics in this article to derail my investigation.

Predatory Gangstalking is also called "terror stalking", "gangstalking", "flash mobbing", "cause stalking," "hate stalking," "multi-stalking, "happy slapping," "workplace bullying, and "covert war."

The reasons why it is employed?

You found the wrong forum on the internet and saw something you should not have seen; you worked for a company and found out some sensitive information; you got on the wrong side of someone who is a member of the vast satanic or luciferian cabal; etc.

What does this have to do with the field of the paranormal? On a macrocosmic level, preternatural beings are the driving force behind activities such as gang stalking. The destruction of a person feeds them. It also fits in with their overall agenda. People who are victims of gangstalking are people who pose a threat to the goals of the dark principalities.

In addition to being stalked by large groups of people, malevolent/demonic entities may be attacking you in the night, and disembodied voices may be harassing you at all hours. Though seemingly unrelated, all of these subjects tend to be connected.

The Goal of Gang Stalking

The expressed goal of Gang Stalking is to silence a victim, drive a victim insane and possibly to the point of suicide, or destroy the victim's reputation and believability as the person will likely be viewed as mentally ill should they complain or report the abuse. Gang Stalking is also used to gather information on individuals as well as force individuals to move or leave an area.

Motivations for the Abuse

Motivations for Gang Stalking vary. Revenge for a real or imagined offense (People in these organizations are quite petty, mirroring their preternatural handlers); false accusations of a "horrible crime" of which the victim has gotten away with (Used to discredit the victim and destroy his/her reputation); silencing a corporate "whistle-blower"; defecting from a cult; a perceived enemy of a group or organization; "knowing too much" are all examples of possible motivations.

Who are the stalkers?

The stalkers, for the most part, appear to be everyday citizens. They wear business suits, push their children in strollers, and shop at the local grocery store. They are encouraged to fit in with the community. Other stalkers are "street thugs" who have been hired to harass and intimidate.

Vast numbers of stalkers seem to have no idea that they are being controlled and used to harass people. Like a vehicle, the entities enter; use/drive them, and then exit. They will be going about their business one moment, then like a switch being flipped, they are your worst enemy. They go about their assignment with an icy hateful demeanor or at times a nearly playful trickster persona. Under it all there is a constant seething malevolence that is palpable.

A gang stalking group is a well-run organization comprised of members who are unidentified as gangstalkers to the outside world. Until a group of gangstalkers turn their face towards someone selected to be one of their targets, members of society do not realize there is such an invisible group.

The practitioners of gang stalking are people who, for the most part, go about their business of daily life appearing like everyone else, except for their activities involved with Gang stalking. Gangstalking activities take priority over everything else in the lives of gangstalkers. Gangstalking is actually a lifestyle for those who participate in it.

Why people participate in Gang Stalking

- Some stalkers are told lies, either positive or negative in nature, in order to gain their participation

- Some stalkers are paid or receive other benefits

- Stalkers belonging to an organization may simply be following orders.

- Some stalkers may use their participation in order to repay a past favor.

- Peer Pressure/ Need to Fit In

- Former stalkers have stated they participated out of fear of becoming the next target should they go against the group.

- Entertainment Value/Thrill of Participation in an Illegal activity.

- Nearly all willing stalkers have a cult/satanic or luciferian group affiliation

- Some are controlled like puppets (flipping the switch)

Who or What is Behind Gang Stalking?

Societies/Fraternities/Orders, Religious Cults and Destructive "New Age" Groups, Corporations, Government Organizations, Military, Concerned" Community Groups/Vigilante Groups, Criminal Organizations, etc.

Ultimately, the source of it all are the malevolent/demonic principalities. They are the true heads of state and corporations in this world.

Writer/Researcher Carissa Conti puts it this way: "The people involved with gang stalking activities may not always be piloting their own wheel, to put it in a straight forward manner. Other "stuff" may be manipulating them, working through them. There definitely seems to me to be a hyper dimensional component to it all – hyper dimensional, as in, non-physical realms and entities, whether positive or negative, that exist outside our own 3rd density physical reality, unconstrained by space-time, manipulating reality and working through people, like puppet masters from behind the scenes."

In 2 Worlds: Gang Stalking vs. Hyper dimensional Matrix Attacks - Updated 3/6/10 in2worlds.net

I highly recommend reading her article at the link above after you read this article. It conveys the otherworldly nature of this form of attack, and gives you the feeling of what it's like to be stalked.

Examples of Gang Stalking Harassment

Flash mobs, Slashed Tires, Threatening Phone Calls, Verbal Assaults by Strangers, Property Damage, Death Threats, "Peeping Toms", Following on Foot or by Vehicle, Bizarre Notes and Drawings Left(usually of a satanic nature), Odd behavior by multiple people in relation to the victim(Everyone makes a certain gesture as they walk by, or mention victim's name while talking on their cellphones), Loitering, Anonymous False Accusations to Friends, Family, and Neighbors, Character Assassination, Smear Campaigns, "Black-Listing", Psychological Abuse, etc.

Tactics used by Stalkers

Tactics of Predatory Gangstalking include highly coordinated surveillance (hidden cameras, conversation bugging in private as well as public), harassment, and psychological, psychosocial, financial, and sometimes physical assaults on an individual by a large group of people who are often strangers to the targeted individual.

Anchoring

Anchoring is a technique employed by stalkers to implant a false motivation or reason behind the stalking, preventing the victim from discovering the truth. In more sinister examples, Anchoring involves the implantation of evidence to persuade the victim some other group or organization is responsible for the abuse.

Baiting

The term "Baiting" is a stalking tactic used to lure a victim into environments, or situations, which cause further problems to the victim. Often "Baiting" involves tricking a victim into committing a crime or unknowingly engaging in an illegal activity such as assault. They trick the victim into thinking an innocent bystander is the culprit of their harassment and try to compel the victim into assaulting the person.

Brighting

Brighting is a term jokingly referred to by stalkers to indicate the practice of repetitive flashing of a car's high-beam headlights. The victim is usually followed and may be "flashed" from either a "tail-gating" vehicle or a passing or on-coming one. "Brighting" also occurs when bright lights are flashed into a victim's home-windows.

Electronic Harassment

Electronic Harassment is the use of technological devices to spy on or cause harm to targeted victims. For example, exposure to a high magnetic field has been shown to induce hallucinations in humans while exposure to intense microwave radiation induces psychotic episodes and causes brain damage.

A frequent form of Electronic Harassment involves beaming a low frequency "hum" or "tone" into a victim's home or general area. Over time, the exposure causes the victim to lose sleep, become agitated, and suffer the effects of prolonged stress. Such tactics are also being used in cases of hostage situations as well as covert government operations. These forms of magnetic fields can be caused by technological or preternatural means.

Ghosting

The term Ghosting refers to the practice of rearranging, or moving, of a victim's home furniture, lawn decorations, desk decorations at work, etc. The purpose of Ghosting is to make a victim question his or her sanity. Ghosting is also designed to make other's question the sanity of the victim, especially if the victim attempts to complain of the abuse.

Mimicry

Mimicry is a specialized form of harassment in which the stalkers publicly imitate every movement made by the victim. This shows an aspect of the trickster. It is a manifestation of the true source of the stalking.

Mobbing

Mobbing is a term that describes "Group Bullying". Of itself, Mobbing is not equated with gang stalking. However, Mobbing may be a tactic used by the perpetrators of gang stalking.

"Flash mobbing" is an event occurring when a targeted individual is spotted in society. Their photo and location are immediately and simultaneously sent out by cell phone text messages and Internet email to all gangstalkers within a certain radius. All gangstalkers then suddenly descend upon the target as a mob.

Noise Campaign

A Noise Campaign is an orchestrated effort to produce stress in a victim through prolonged exposure to significant noise levels. A Noise Campaign can range from multiple neighbors routinely playing loud music, individual stalkers with air-horns or fireworks, or organized "repair work" that involves a high level of noise.

Sensitization

Sensitization is a psychological term referring to the forced association between a stimuli and a corresponding reaction. Stalkers use Sensitization to psychologically abuse a victim. For example, if a stalker constantly harasses a victim while wearing a blue baseball cap, then overtime the victim will begin to believe anyone wearing a blue baseball cap is a stalker and is coming to harass. Sensitization undoubtedly creates an extreme level of fear in a victim, in direct fulfillment of the intentions of the stalkers.

Street Theater

Street Theater is a term used to describe the odd-actions and behaviors that stalkers do in public, in an attempt to rile the victim. Such behavior often borders on the extremely bizarre and is aimed at a blurring of the boundaries between reality and fantasy in the minds of the victims.

Examples of Street Theater: Baiting, Brighting, Color Harassment, Convey, Directed Conversation, Ghosting, Mimicry, Noise Campaigns, etc.

Variations of Gang Stalking

Gas-Lighting

The term Gas-Lighting originates from the 1944 film Gaslight. In the movie, the character of Gregory Anton, played by actor Charles Boyer, attempts to drive the character Pauline, played by actress Ingrid Bergman, insane. The phrase Gas-Lighting has come to mean similar actions and behavior, as used in the film, against a victim.

"The White Glove treatment"

"The White Glove Treatment" is slang for Gang Stalking as jokingly referred to by the Masons who either participate in, or know of its usage against so-called "Enemies of Freemasonry".

[Note: The phrase "The White Glove Treatment" is not limited to Freemasonry, as members of similar groups and organizations often practice Gang Stalking.)

Ritual Gangstalking

Reportedly performed by cults and secret organizations such as Satanists, secret orders of free masons, some voodoo /obeah practitioners and Scientologists as a means of control or for punishment. "Ritual abuse" is incorporated into "ritual gangstalking" patterns in this case.

Cyber Gangstalking

A situation where an individual is identified to be a target through Internet participation and their personal identity is determined. Cyber

Gangstalkers pierce the electronic veil and are able to enter into the real life of a person, creating all forms of harassment, theft, psychological operations including slander and libel, and able to commit crimes against the targeted individual.

Sent from Yahoo Mail on Android

November 17, 2014 – This is the third email to Derrick Robinson asking for Robinson's email address to send him a video as he had told others on the 14th during first contact:

Subject: Volunteer

From: Myron May

To: Derrick Robinson

Can I have your mailing address?

November 19, 2014 – May email sent to several individuals detailing certified mail packages are in route with images of postal receipts.

November 20, 2014 - At 12:25 a.m., police receive reports of a gunman at Florida State's Strozier Library. Three people are shot. May is gunned down by 12:27 a.m., police say. (Source: Tampa Bay Times. Link Below:

http://www.tampabay.com/news/education/college/officials-man-accused-of-shooting-three-students-at-fsu-library-attended/2207287?ref=hihidnews)

They got there so fast it almost appears that they knew he was coming.

CHAPTER FIVE
The Renee Pittman Connection

Beginning on November 14, 2019, I received the first Facebook in box message from an attorney identified as being named Myron May. The day after the shooting, the operation targeting me watched me as I tried to copy the messages to save on my computer. Believe it or not, right before my eyes, our interactions were immediately deleted and May's Facebook page was shut down at that same time. The message "Sorry, No Longer Available" flashed on my laptop computer screen related to his Facebook page.

Later, I would learn that his Facebook account was taken down for a day, on November 21, 2014, and when it was put back up on approximately November 22, 2014, however, all of his comments to me had been removed by some agency. As a result, my responses were still visible in the messaging only at that time. In the place of Myron May's responses back to me it now read:

This message has been temporarily removed until we can verify the sender's account.

As a result, I still copied and pasted my responses to him below and attempted to piece together from memory my very first interaction with Myron May on Facebook beginning on November 14, 2014.

About a month later, as I embarked on this project, and went back to view my Facebook message responses, for accuracy, I would learn that now even my responses had also been completely deleted back to Myron May also. As a result, it would appear that there was no evidence of Myron May's first contact with me whatsoever on Facebook.

However, I remembered that at the request of the NBC reporter on November 21, 2014, and also a request for a guest appearance on "The Pete Santilli Show" that I had compiled the voicemails, emails, text, and Facebook messages together and email both the information. As a result, I was again able to retrieve only my questions and responses to Myron May.

FACEBOOK NOVEMBER 14, 2014

November 14

11/14, 10:22 a.m.

MYRON MAY

This message has been temporarily removed until we can verify the sender's account.

MYRON MAY

This message has been temporarily removed until we can verify the sender's account.

Shown above are the initial Facebook interaction between Myron May and myself where he simply introduced himself.

November 14

11/14, 11:17pm

MYRON MAY

This message has been temporarily removed until we can verify the sender's account.

Here is one of the first messages where I responded after May asked me if I had an attorney for the Pro Se case pending in Federal District Court. The case is related to intense physical torture which materialized as ongoing efforts around me to get me to stop book publications using coercive physical torture via Directed Energy Weapon attacks, 24 hour a day, 7 days a week in an ongoing technological harassment effort from the operation center of those assigned to me

RENEE PITTMAN M.

Yes. I file Pro Se and the judge is listening. It has not been dismissed yet which is a good sign after the US Attorney moved for dismissal Lack of Subject Matter Jurisdiction. I recently did a second Motion for Appointment of Counsel. It was denied "Without Prejudice" which is a good thing I think. I read that typically attorneys are not appointed in Civil Complaints. However, twice they have denied it "Without Prejudice."

11/14, 11:21pm

RENEE PITTMAN M.

I also can create reasonable doubt related to the beam deteriorating my body. I have been tracking by repeated x-rays every two months the deterioration. And they made the mistake of advancing illness which is out of the ordinary and unexplained by energy weapon deterioration in medical studies. I think I can get them but there is no way I can do it without counsel.

11/14, 11:22 p.m.

RENEE PITTMAN M.

If I have to represent myself, and have no other choice, I will. But you know they say that representing self factually equates to having a fool for a client.

Saturday

11/15, 4:08pm

MYRON MAY

This message has been temporarily removed until we can verify the sender's account.

Above is Myron May's request for my address.

RENEE PITTMAN M.

I asked him what the video was about but he said that he did not want to say. In hindsight, I feel that he had planned to not be among the living when the video was received.

11/15, 4:09pm

RENEE PITTMAN M.

I provided May with my address then asked "What's it about?"

11/15, 4:32pm

RENEE PITTMAN M.

I asked again, "What is it about?" for the second time due to no response from him.

11/15, 4:32pm

MYRON MAY

This message has been temporarily removed until we can verify the sender's account.

Unsure what he said here. I do recall trying to pinpoint him about what the video is about? When he would not say, I began to get suspicious of someone I had given my address to.

11/15, 4:32pm

RENEE PITTMAN M.

Do you want it published?

11/15, 4:33pm

MYRON MAY

This message has been temporarily removed until we can verify the sender's account.

I remember he simply responded "Yes" here.

11/15, 4:34 p.m.

RENEE PITTMAN M.

Before I ask any more questions, I will wait for the video. Take care and hang in there!

Sunday

11/16, 11:12 a.m.

MYRON MAY

This message has been temporarily removed until we can verify the sender's account.

    He mentioned he was sending only a video at the time by Certified Mail. We spoke briefly about an incident recently with me where Certified letters I had taken to the Post Office were returned to me without any effort whatsoever for mailing.

11/16, 11:13am

MYRON MAY

Here is where he asked me do I get hit with Direct Energy Weapons.

RENEE PITTMAN M.

Around the clock. Two hips destroyed by the beam and two knees being destroyed while I sleep each night.

Can't you send me the video by email?

11/16, 11:20am

Below are messages I kept sending. When I did not get a response, from May, this furthered my suspicion, initially, that something else was going on here of which I had not a clue and did not want to be involved in. Typically, messages are immediate exchanges in conversation back and forth between two parties.

When I sent the messages below, and did not get a response, I blocked Myron May on Facebook in hope of this erasing hopefully my address on November 16, 2014.

RENEE PITTMAN M,

With a good attorney I have a chance of proving the deterioration. Why do attorneys not want to take on these cases?

Hello! I must admit I get cautious when I send a return message and get no response from someone who initiated the conversation with me. What's up with this?

I also regret giving out my address. I don't want some deranged person to show up at my door.

Hello!

Okay. You are deleted. I have found that black men, those targeting me, and others to be some of the most mind-controlled people on this planet and vicious. Lose my address. I have got my own problems.

Twice you have not responded when I asked you a question. I just have no time to play around. I can't help you.

11/16, 11:33am

RENEE PITTMAN M.

Because you do not respond, I feel like you can't be trusted. Again, I regret giving my address to a complete stranger who could show up at my door. If you show up at my door, I can guarantee that the police will arrest you just to get rid of you along with all other black people! Bye

NOTE: I know this sounds harsh but I just did not know how to take him and wanted to scare him if he was a real nutcase from showing up at my door. I would later unblocked him, after his requests for me to do so sent to my Twitter account and also I later learned on November 16, 2014 also posted on the Targeted Individuals International Facebook group page.

Monday

11/17, 2:30pm

MYRON MAY

This message has been temporarily removed until we can verify the sender's account.

As shown on Twitter messaging, after I unblocked Myron May, I asked for his phone number to hear a human voice at least. He gave it to me and we spoke for a while where he explained his targeting.

Much of the time was spent with me trying to explain the dynamic of what is happening today. During our phone conversation Myron mentioned that he felt that nanotechnology was the key to the tracking and torture.

I later sent him the links below in box messaged on Facebook to him in my attempt to explain to him my perception of the program as being part of the Total Information Awareness Program, Biometric Surveillance using a Bio-coded energy weapon system.

Tuesday

11/18, 9:41pm

RENEE PITTMAN M.

I don't believe in nanotechnology. This is what I think is happening

(1) http://www.epic.org/privacy/profiling/tia/

EPIC Terrorism (Total) Information Awareness Page epic.org

EPIC Urges Scrutiny of Proposed Federal Profiling Agency. In a letter (pdf) to a House subcommittee, EPIC urged careful scrutiny of the Department of Homeland Security's proposed Office of Screening Coordination and Operations. This office would oversee vast databases of digital fingerprints and pho…

11/18, 9:43pm

RENEE PITTMAN M.

(2) COMBINED WITH THE MILITARY BIOMETRIC "LOCAL POPULATION CONTROL" PROGRAM. This would explain the black men voices who are possibly US Army personnel.

LINK: http://www.eis.army.mil/programs/biometrics

PEO EIS – The Army's Technology Leader

www.eis.army.milOrganizations within PM DOD Biometrics includes: Biometrics Enabling Capability (BEC) BEC consists of the Next

Generation-Automated Biometric Identification System, also known as (NG-ABIS), is the central, authoritative, multi-modal biometric data repository.

11/18, 9:44pm

RENEE PITTMAN M.

People believe chemtrails are dispersing nanotechnology. This is what I believe regarding chemtrails.

(3) http://rense.com/general79/chem.htm

What chemtrails really are - rense.com

11/18, 9:48pm

RENEE PITTMAN M.

(4) https://plus.google.com/+AuthorReneePittmanM/posts/R9aW3sRupVG

BIOMETRIC SURVEILLANCE Biometric surveillance is any technology that measures... plus.google.com

BIOMETRIC SURVEILLANCE Biometric surveillance is any technology that measures and analyzes human physical and/or behavioral characteristics for... - Author Renee Pittman M. - Google+

11/18, 9:49pm

RENEE PITTMAN M.

(5)

http://educateyourself.org/cn/flemingshockingmenacesatellitesurveillance14jul01.shtml

The Shocking Menace of Satellite Surveillance by John Fleming (July 14, 2001)

11/18, 9:49pm

RENEE PITTMAN M.

Hang in there!

p.s. Drones will also step up to the plate in Directed Energy Weapon attacks. If you wake being tortured by Directed Energy Weapons look into the sky for the flashing lights of a Drone or two. You can see them in the wee hours of the night!

11/18, 9:52pm

RENEE PITTMAN M.

http://www.washingtontimes.com/news/2012/feb/7/coming-to-a-sky-near-you/?page=all

Drones over U.S. get OK by Congress

www.washingtontimes.com

Look! Up in the sky! Is it a bird? Is it a plane? It's ... a drone, and it's watching you. That's what privacy advocates fear from a bill Congress passed this week to make it easier for the government to fly unmanned spy planes in U.S. airspace.

11/18, 10:03pm

RENEE PITTMAN M.

FYI - You might not even have ADD

https://www.youtube.com/watch?v=gvdBSSUviys

Psychiatry: An Industry of Death (FULL VERSION)

Full credit for this video goes to CCHR. I have no connection with Scientology, religious or otherwise. I am however a former victim of criminal psychiatric writes the person who posted the history on You Tube.

11/18, 10:04pm

RENEE PITTMAN M.

Hope you're still among the living Champ!

Wednesday

11/19, 3:52pm

MYRON MAY

Deleted response after his account was closed down then put back up.

This comment was said joking with Myron May as an attempt to lessen his mind set of something many targets experience. Again, I did not believe for one minute that he actually was suicidal, nor am I qualified to make this determination. Later even a visit to a psychologist reported him as fine. This was especially true based on his calm demeanor. With the exception of my deleting him and his urgency for me to unblock him, being the only thing that was odd or aggressive to me.

11/19, 3:53pm

MYRON MAY

Deleted Response by some agency

11/19, 3:55pm

Deleted Response by some agency

MYRON MAY

The above was Myron May saying he sent postal receipts to me to prove the certified mailing.

November 21, 2014 – 9:30 a.m.

Just one day after the FSU shooting, the Postal Inspector, an FBI agent and a local Sheriff knocked on my door. In the postal inspector's hand was the certified mailing Myron May had reported to be sending to me. I was told by the inspector that the "Certified" mailing had been left sitting on my front porch by my regular postal carrier.

NOTE: Incidentally, on December 1, 2014, when my postal carrier knocked to return two certified mailings unclaimed by neighbors I sent out telling neighbor's why I was factually being targeted, that I was not on drugs or mentally disturbed, I asked him if he recalled leaving a certified package on my porch on the morning of Friday, the 21st of November. He said he was working that day and he had not.

The three entered my home, with the inspector showing his official badge, but nothing from the man identified as being an FBI agent was shown except an introduction as being an FBI agent. The female sheriff officer introduced herself, and immediately explained to me that she was there for quality assurance, so to speak.

The inspector and I took seats at my dining room table while the female sheriff, stood to my right, and the FBI intentionally positioned himself directly behind me, standing as I sat, and as I spoke with the inspector, now out of my sight.

I was first asked did I know Myron May. I replied yes. I was then asked how I knew him. I explained that he had contacted me on Facebook initially on November 14, 2014 just six days prior.

I was asked did I know of the recent FSU shooting rampage by Myron May. I told them yes, but that I had not been informed about his role until an associate contacted me from Northern California, who was also listed as a recipient of one of the packages. At that time, we were both in shock. That was around 3:30 p.m. November 20, 2014, I explained. I also stated that I was working on a project and that I don't typically watch TV much overall.

I was asked did I know that a certified package was being sent to my home, and also being sent to numerous individuals by Myron May. I told them yes. May had informed me of this through various messages.

I told the inspector, that I had received numerous text messages from May on November 19th late afternoon while running errands. He stated via text that he was sending out four certified packages I thought. We were texting back and forth while was I was in a store cashier line and I got out of the line to text him back.

I told them that May reported to me that after he made the official mailings at the post office, he immediately checked tracking information online. He said that three mailings, one sent to my associate up North, and two others were showing via tracking information that they were in route, however May stated that the mailing he had sent specifically to me was MIA with no tracking information of progress whatsoever. He felt it was the result of Post Office tampering.

NOTE: As shown, a large portion of the Gmail attachments were images May has taken of license, plates, of vehicles cutting him off, and several people he said had actually followed him into the post office while he was in line to mail the packages. He also, as shown earlier, by the attachment listings, provided images of an individual he said had stepped up to the counter, and whispered into the postal clerk ear something negative about him he felt. Because of this he indicated possible post office inappropriateness and that this specific incident of influencing must have impacted his mailing designated specifically addressed mail to me.

Within the organized / Gang Stalking dynamic, many, many targets report being targeted everywhere and followed wherever they go. I personally have had several of these types of similar experiences where those sitting in the operation center, watching me in real time, would call ahead to a restaurant, for example, of which they knew I was heading to through thought deciphering technology, and when I arrived, there was a cold chill directed at me, and in some cases unexplained rudeness. In a few incidences, the person I was eating with their food would arrive piping hot and mine cold as if out of the refrigerator. Needless to say, there was no tip from me and I even requested and received a refund on one occasion.

The general public termed InfraGard is heavily connected to mobilize organized stalking via community businesses as I mentioned earlier. There is also a list of restaurant participants other tracked targets report as having similar problems with.

InfraGard is the mobilization effort of organized stalking specific to the FBI counter terrorism division. Today also, state and local police departments also have technologically advanced counter-terrorism divisions deemed necessary in the new Militarized Police State. However, again, ALL agencies today fall under one Intel umbrella and it is believed that local police are going to be now escalated to the Federal level.

The InfraGard website states, at infragard.org:

"The InfraGard is a partnership between the FBI and the private sector." It is an association who represent businesses, academic institutions, state and local law enforcement agencies, and other participants dedicated to sharing information and intelligence to prevent hostile acts against the U.S."

Many harassed targeted individuals in ongoing reports, report the stalking also uses toxic fumes spread into their residences.

Because the text messages were lengthy, while I was at the store, May decided to phone me to explain himself and we spoke briefly before I left the store with my telling him I would check my computer, and call him to see if I had the emailed tracking information he had sent to me when I got home. I told the inspector as the other two listened intently.

I did not get home until around 5:30 p.m. due to stopping to eat, I immediately checked my email address and phoned May stating that I did see tracking information for not just four but several notifications coming from a Florida Post Office. We spoke briefly.

I was confused because he said initially told me there were only four packages. May explained that he had intentionally sent all of the notifications to me as proof of mailing to the other recipients also before we got off the phone. Afterwards, May began to inundate me with emailed images of the receipts, to include on Facebook, of images he had taken with his cell phone of the receipts.

I told the inspector that during a previous conversation with May, on or about the 15th of November, right after his initial contact with me, on the 14th on Facebook we discussed via Facebook messaging his sending a video to me. Coming from a complete stranger this quickly aroused my curiosity and my wanting to know what it was about or contained. I told them that the video, as being a suicide video, was the farthest thing from my mind and on the 19th my associate up North and I joked that she hoped the package held money.

The next day after May's initial contact with me via Facebook messages, on November, 15, 2014, I realized that under the circumstances, surrounding my life, that I had given a complete stranger my home address. I also was cautioned because during our Facebook messaging there was unusually long gaps and pauses in the back and forth Facebook messaging exchange with him. Under the circumstances, and because it happened so frequently, I felt that it was not unrealistic that he could have been possibly getting coached on what to say to me. I told the inspector that I am heavily targeted and many people have been used to include family, friends, and everyone under the Sun, which also includes strangers attempting to ingratiate themselves with me in ongoing efforts for now eight years. I told him it is related today due to the books I have written and a hope to neutralize me. The sheriff asked me the name of the books and I told her.

On or about November 15, 2014, I began to feel it was a mistake to have given my address to someone I just met the day before and felt that the only way to hopefully delete my address was to block him on Facebook which I did. I texted him also that I did not trust him.

When I did, May sent the following messages via Twitter.

The first tweet as at 1:23 p.m., with the other tweet exactly the same, as shown below tweeted at 1:24 p.m., on November 16, 2014:

The May Firm@TheMayLawFirm

@ReneePitttman

Renee, Please add me back on Facebook. I really need you. I just saw your Facebook message.

The May Firm@TheMayLawFirm

@ReneePittman

Renee, Please add me back on Facebook. I really need you. I just saw your Facebook message.

Again, I explained I still did not understand why there were such long breaks with him until he said that he was being attacked as he typed responses.

The May Firm @TheMayLawFirm Nov 17

@ReneePittman

and they are attacking as I type

12:39 PM - 17 Nov

Reply to @TheMayLawFirm

Renee PittmanM @ReneePittmanM Nov 17

@TheMayLawFirm

I unblocked you. Inbox me your phone number. You are blessed and will prove vital to exposing this effort.

The May Firm @TheMayLawFirm Nov 17

@ReneePittmanM

I'm legit.

On November 16, 2014, May also posted in the Targeted Individuals International Facebook group, along with Twitter.

Below is a sequence of message he first posted related to Class Action Lawsuit interest as shown within this group's page on November 14, 2014 of which my associate up North was spearheading via another Facebook Group she created named TI Class Action Lawsuit. Here are the posts on the Targeted Individuals International group Facebook page:

MYRON MAY

November 14 at 10:23am

I am a lawyer. Are there any other lawyers on here? Perhaps we can put together a legal team to get something going. If you're out there, respond. There is strength in numbers.

MYRON MAY

November 15 at 10:17am

Per Name Deleted: I have just created a group (tonight) for every TI who wants to pursue a class action lawsuit ... let me know if I should add any of you.

If you want to join a class action, contact Name Deleted.

MYRON MAY

November 16 at 1:19pm

Dear Renee Pittman Mitchell:

I just saw your message. Please add me back. I literally just saw your message. I REALLY NEED you!!!

Because May was an attorney I thought he was a God send I told the Postal Inspector.

I explained to him why May, felt possible post office tampering after law enforcement influence during the mailings at the post office.

When I asked for May's phone number I told him he would prove to be a blessing to the targeted individual community due to much needed legal help and expertise, I firmly believed and hoped he would be.

During a subsequent conversation with May, about the 16th after I have unblocked him on Facebook, and after he had given me his phone number. I mentioned to May, that the Post Office, not mailing something intentionally that was certified would not surprise me at all. I told him that a week prior on November 3, 2014, after recently moving where I currently live, law enforcement had been going through my new neighborhood attempting to organize an organized stalking effort by telling neighbor's falsehoods in order to get them to participate.

After realizing this reality, and as a result of it, I sent out several certified mailings to neighbors in homes around mine disputing this wrongful disinformation. I told him that since 2008, I have been aware that the new paradigm via law enforcement which entails "Technological" harassment, Directed Energy Weapon attacks, including portable technology coming from neighbor's residences in every location I have lived. Many targets have posted their experiences everywhere as being similar also. Also, from my past experiences several locations had been set up to watch me inside my home, by

The Targeting of Myron May

USAF personnel using microwave technology, and law enforcement, under the guise of me using drugs, using patented "see thru the wall infrared technology" which led also to training of neighbors on portable versions of Directed Energy Weapons which can carry also the operator's voice by radio frequency radio waves.

At that point the Postal Inspector's phone rang, and he excused himself and stepped outside to talk privately.

While outside, the female sheriff asked me to explain to her what electromagnetic weapons are? I told her they are radio frequency weapons. She then shook her head in recognition and asked what the range of the beam is. I told her for the Active Denial System, 3.2 miles. Satisfied with my answer, she then was quiet as the inspector returned.

They asked me if I had any other contact with Myron May regarding the mailings. I told them that I had also the voicemail messages, along with the texts and emails. I excused myself to go upstairs and get my lap top and phone.

Believing this also yet another opportunity to educate the public about what is happening today, I first showed the inspector my website and began to explain the targeted individual program, laws approving it, and numerous mind control official patents listed on the site.

At that point, the inspector's phone rang for the second time, and he again excused himself, and stepped outside to speak with the caller. I heard him saying, "Yes Sir. Yes, sir."

While outside the person representing himself as an FBI agent remained positioned behind me. I then focused my attention on the female sheriff. In effort to show that May likely was not crazy in light of what many know is a day to day reality in their lives, I attempted to show her the mind control patents, asking her to step forward and take a look at my website, bigbrotherwatchingus.com. She was a little apprehensive and first looked at the FBI agent of whom I ignored.

I told her the "hearing voices" effect which Myron May reported and what literally thousands of intelligent people are reporting today can actually be accomplished by technology the DOD officially calls the "Voice of God" which is also called Voice to Skull. As I was doing this, the inspector returned and the FBI agent stepped forward saying someone to the effect that he was crazy too. I looked at him and said I was told, "Crazy people do not think their crazy, so if you are saying you are crazy, it means you are completely sane.

NOTE: This was said by a VA doctor.

I perceived the effort of this man as a hopeful act of intimidation. He smiled a crooked smile then repositioned himself again behind me and the inspector continued the line of questioning.

However, the inspector now had a pen and paper in hand and stated that he was simply taking notes. I stopped him there telling him, looking around at the FBI agent, and the sheriff that when they knocked on the door I had been factually upstairs working on Myron May information related to my interactions with him of which I had just forwarded to an NBC reporter to include the voicemail recordings.

I told them that I would forward the emails on to him which detail all interactions with Myron May along with the certified mail images. With this said, the meeting began to wrap up with him leaving his business card with his email address on it for me to forward the information.

Before he stood, he asked, "Do targeted individual typically do these types of things, meaning go literally postal?" "No, I responded, with close to 300,000 being targeted today at one level or another in this program, it is rare," I told him. As they reached the door, I said, "Not to sound like a conspiracy theorist, I would say, "False Flag" and gun control. With that said, I shook each of their hands and the official team left.

At some point, I believe when they first arrived, I also mentioned focusing on the FBI agent that due to the ongoing extreme harassment effort around me, that I had also sought counsel from Gloria Allred, just in case, someone tried to link me to Myron May's tragic decision as being involved.

Later that afternoon, November 21, 2014, nearing evening, a Reuter's reporter also contacted me via Twitter providing her phone number and asking me to call her as shown below via Twitter. As a result, likely of NBC article:

Mark Wallheiser / Getty Images

NBC article excerpt:

FSU Shooter Myron May Left Message:

"I Do Not Want to Die in Vain"

NBC (Universal)

By Tracy Connor

Hours before he opened fire at the Florida State University library, lawyer Myron May left a desperate voicemail for an acquaintance with this plea: "I do not want to die in vain."

The message was part of a flurry of emails, texts and phone calls in which the former prosecutor laid bare his torment: He believed government "stalkers" were harassing him and using a "direct energy weapon" to hurt him. He said that he had sent packages to 10 people that would "expose" what he thought were happening to him.

NOTE: Excerpt from interview with me by Tracy Connor

Twitter contact by Reuters reporter:

@ReneePittman

I am a Reuters reporter hoping to speak to you re Myron May. Can you pls call at Phone Number Deleted

Below is the excerpt from the Reuters phone interview later published in Reuters. I also provided this reporter with a copy of May's suicide letter.

Letters Detail Thoughts of Gunman Killed At Florida University, Excerpt - By Bill Cotterell

Crime scene tape is seen in front of the library at Florida State University, in Tallahassee, Florida, November 20, 2014.

Credit: Reuters/Bill Cottrell

"One was intercepted by the FBI's Houston office, a spokeswoman said, noting it contained nothing hazardous.

In California, authorities on Friday morning removed a package from the porch of Renee Pittman, she told Reuters.

The Targeting of Myron May

She is the author of several books and a website on what she calls "mind control technology," detailing a covert effort to target people with radio frequency technology.

May reached out to her one week ago, she said, believing he was a target.

"He said, 'I just don't want to live my life like this,'" according to Pittman.

She said her package contained a letter and a thumb drive. May told her that she could expect eight more mailings.

She did not thoroughly read her note before turning it over to authorities. Another recipient, however, forwarded her a typed letter that was addressed to her and several others, images of which Mitchell sent to Reuters by email.

The letter, signed by May, explained that he had told no one about his plans and that his goal was to bring media attention "to the plight of targeted individuals."

"I would like to make a sincere plea to you not to let my personal story die," it read...

(Reporting by Bill Cotterell; Writing and additional reporting by Letitia Stein; Editing by Lisa Von Ahn, Jim Loney and Eric Beech)

As God as my witnessed I had not been even remotely expecting that this soft spoken, obviously brilliant, highly intelligent young attorney would do what he felt his option, after great and penetrating losses of his life by human monsters working these operations.

During a televised interview with ABC, Tampa Bay, News Reporter Courtney Robinson I made this clear as the truth:

Woman recounts voicemails from FSU gunman

Courtney Robinson, 12:16 a.m. EST November 25, 2014

LINK:

http://www.wtsp.com/story/news/local/2014/11/24/fsu-shooter-left-voicemails/70072098/

Expert: Myron May "He had no exit strategy."

Updated: Wed 1:01 AM, Nov 26, 2014

LINK: http://www.wctv.tv/home/headlines/283926011.html

In May's beaten down to a pulp confusion, it appears he believed he was doing the honorable thing by sacrificing himself. However, his scheme in reality left a negative impression of targeted individuals. It says that, no matter how highly educated you are, you could snap and hurt self and others.

I would later read in a news report that the Taunton's reported that if you put 100 people in a room together with Myron May, May would be the last person, if at all; they would suspect would do something like this at all or even remotely close.

And I, thousands of miles away, had not a clue of a person basically a stranger to me and inundating me with messages.

Voicemails sent to my cellular phone which I also gave to NBC and also the united states postal inspector, with FBI and local sheriff present:

NOVEMBER 19, 2014:

- 6:19 p.m. - Renee I really need you, too, to to answer the phone. This is, this is, not a game, this is not a joke. Renee, I need you to answer the phone. I need you to answer it, ahh now or answer my text messages. I do not need you to bail on me right now. This is a terrible time for you to bail on me right now I really need you. Please do not do that, please.

- 6:22 p.m. - Renee look, I'm getting hit with Direct Energy Weapons, right now as we speak. I have nine return receipts coming to your address. I need you to not be paranoid and calm the Hell down because I do not want to die in vain. Okay, like seriously

- 6:42 p.m. – Renee listen to me. I need you really bad right now. I am currently being cooked in my chest. I devised a scheme. Aah, where I was gonna expose this once and for all and I really need you. Like I really, really, need you and your paranoia is kind of messing me over and I do not want to die in vain. Renee, I need you to please pick up the phone, check your email, do something because otherwise I'm screwed.

Below is my text correspondence with Myron May the morning of November 19th 2014. I had hoped to get a response from him of whether he would represent me in my Pro Se case, and also hopefully, ultimately in the Class Action lawsuit of which I was also part of the originating thought to attempt via ongoing monitoring of my efforts.

November 19, 2014 - Wednesday 9:00 AM

RENEE PITTMAN M.

I am willing to bet that the torture increased after talking with me as a warning to not get involved. Am I right or wrong?

NOTE: I had learned that any request for legal assistance and any exposure efforts typically amp up Directed Energy Weapon attacks as a message to stop exposure or else.

November 19, 2014 - Wednesday 2:48 PM

MYRON MAY

It's about the same

RENEE PITTMAN M.

"I think it is the same black guys based on something they said when we got off the line. They said they would have to leave both of us alone with a judge reviewing the case. I sent you some reading material on FBI.

NOTE: Myron had reported to me that it was a group of black men targeting him in the week before the rampage and about the time that he first connected with me. He may have come under their radar as have everyone around me after contact with me by this specific group. This specific group of men I have reported several times to be vicious beyond belief believing that my book publications will eventually cost them their jobs as the foundation.

RENEE PITTMAN M.

Hang in there, Champ!

Oops FB not FBI. That would mean the Army Operation Center in Virginia who spearheads the biometric tracking program if it is the same group of Expletive Removed

RENEE PITTMAN M.

Anyway. Don't feel pressured to represent me. Sorry I asked. God will send the right person or I will fly solo. I am thinking that might help on appeal. Typically, courts don't appoint for civil matters but I read can at times at the Appeal level. Is this true?

Plus, they might zap you during cross examination. (Smile)

MYRON MAY

Yes, that's true. You can sometimes get a court to appointment at the appeal level.

Please don't take offense or think I am brushing you off. I'm not committing yet because I'm ironing out a lot of things on my end.

You didn't say anything about my text regarding the certified mail not showing up in tracking.

MYRON MAY

Renee,

I sent ten packages priority mail, certified mail, return receipt requested today, November 19, 2014. Nine of the return receipts are going to you because I anticipate I will be gone at that point. The other return receipt, which is the package being mailed to you, is going to Derrick Robinson from FFCHS. Right now, I am being direct hit as I type this. I hope you get it.

Don't overreact to that. Just take a deep breath.

NOTE: It was after the above revelation that I factually decided to have no further contact with this stranger and several mixed confused messages.

November 19, 2014 - Wednesday 6:13 p.m.

RENEE PITTMAN M.

It appears you are being used to play on my humanity side with your suicide crap. You never…

MYRON MAY

No, Renee. Stop being paranoid. I'm a genuine guy. And please save that email offline so it won't get deleted. I'm still getting hit right now

MYRON MAY

Also, I have another email. It's a Gmail account. I am telling you this because I can't send this file through yahoo because of the size limits.

NOTE: Myron May then text me the Gmail email address.

RENEE PITTMAN M.

You never mailed anything and you know it. I guess the Expletive Removed thought they had a handle on me through my sympathy. You are a shill. You say you want to kill yourself, go for it. That is your business and not mine. Sorry kid, I can't save you.

NOTE: Again, I did not trust Myron May because this group of African American men who have tried everything with me.

If any of you have read my third book in the "Mind Control Technology" book series, I detail also in Book III, "Covert Technological Murder," the story of Jeremiah Ivie who was placed in a very similar position as Myron May. Ivie through extreme physical torture was told that if he stabbed his parents that those identifying themselves as a military operation would stop torturing him. They likely admitted to who they really were because they know that typically, no one believes targeted individuals, or the reality of what is happening today or the efforts of these operations so heinous.

The Targeting of Myron May

There is Nothing Honorable or Patriotic in This Operation

For the record, again, I have also have noted several times that it is men of ALL races, targeting people of all races, and not just the group around me and it appears later Myron May or who surely knew what was happening around him. I believe that May was first targeted by racially motivated cops in Texas, New Mexico, etc., and by the time he made it back to Florida May was being groomed as a patsy possibly for the ongoing gun control "Open Carry" debate in Florida.

Today, in hindsight, I have asked myself, could the thought of legal representation and a real attorney joining my court effort, and a possible Class Action Lawsuit, have escalated the attacks and effort around Myron May to silence him permanently but not before making him a useful tool for globalization? Again, this is why I asked him, via text had the torture increased when I first texted him the morning of November 19, 2014.

However, it not only appeared that a law enforcement effort could have erupted around me to connect me as being involved, but later I came under full attack in an overall discrediting effort against me from two individuals on Facebook, many reports as shady and one possibly as a disinformation agent. Both of these two I had never heard of before. Many who came to my defense, reported both either opportunist for one, and the other a self-described blogger seeking limelight.

The first is a young man who repeatedly in boxed me on Facebook after my name was revealed by the NBC article. It appears he intensely felt that I factually had information he could use to promote his cause connected to his extreme targeting.

The fact is that both Aaron Alexis and Myron May, were virtually unknown to the Targeted Individual community of many activist targeted, with some targeted for well over 20 years until these two appeared out of nowhere.

As the information trickled in after the shooting, and synchronicities emerged, it was revealed that the funeral service preparations were being coordinated by a family owned business, unbelievably to me called the Pittman Christian Memorial. It later was changed to the Christian Memorial Chapel. As a result, the blogger began insinuating a possible connection to me in some way.

In a November 24, 2015 article in Beforeitsnews.com written by Glenn Canady, I was also labelled me as a covert Federal Agent who possibly took part in engineering the shooting. The article is entitled: The article was entitled "FSU Shooter said Government Tortured Him with Directed Energy Weapons! / Alternative

I was also contacted by a targeted individual who said, "I did not want to say anything about this at first. But remember when I told you about someone in Australia contacting me via Facebook, asking me to come there and live, in hopes of getting my targeting to stop by getting out of the country? Well, he said that he lives in Byron Bay in Australia." Byron Bay is actually a location in Australia.

After this I was done. I decided I had enough and immediately deleted people who I felt may or may not be involved, in what now appeared as intense PSY OPS to confuse me.

Someone else had also posted a similarity between Aaron Alexis and Myron May of which was interesting shown below:

Myron May and Aaron Alexis were both mass shooters. Which explains the similarity in their names?

1) Both names have the same letter for the first letter of the first and last names.

2) Both first names have "ron" within them.

3) Both last names are also feminine first names, (Alexis and May).

4) Both names, if first and last names are switched, are also women's names, (Alexis Aaron and May Myron).

Yet another synchronicity.

Admittedly, the Pittman Christian Memorial was uncanny.

When I saw this, admittedly, I was greatly surprised at the coincidence so much to the point that I actually contacted an attorney friend of Mays' who had gone to school with May. He had also attended the funeral. He assured me that in the one stop light town of which May was raised, that is was just that, a coincidence and a fact of thousands having the Pittman family surname across the United States of whom I was not related to.

The blogger, who contacted me after the May incident would later write:

The website for the Pittman Funeral Services (Christian Memorial Chapel) in Graceville, Florida, goes online. Its mailing address is listed as a post office box. The website is where May's obituary would be posted after his death.

The funeral service and chapel are owned by the Pittman family, and has been in business for 52 years.

An online records search for this business only comes up with a record for Pittman Funeral Homes, Inc. in Naples, Florida, which dissolved in 1988.

Is this family business connected to Renee Pittman Mitchell? Why does it not come up as a registered business if it's still in operation? Why is the mailing address listed as a post office box?

In fact, I decided I would like to know myself. There are two funeral service providers located near Wewahitchka and one of the owner's family names just happens to be Pittman.

Curious I phoned the Pittman Christian Memorial Chapel. I was told by an employee, who answered the phone that the family was at funeral services and would not be available until later that evening. I asked him had the funeral actually been held at the Pittman Chapel.

The employee stated that it had not. He stated that a family member had contacted the business now called, Christian Memorial Chapel and requested that the business publish what they call a "Program" which he said is what they call the obituary. I asked him several times could he tell me the name of the actual location of where the funeral services were held. He then stated that as an employee he could not provide that information. He again told me to call back to speak with the owner who is listed as Mr. Donald Pittman.

This funeral service provider is located, in Graceville, Florida, is African American owned and about 74 miles from Wewahitchka, Florida. The fact that it is African American owned contributed likely to its selection.

I call back later during which time I was able to actually speak directly with Mr. Donald Pittman the owner. I asked him for a letter, substantiating no connection with me as a family member, the evening of December 4, 2014. I wanted to insure it was known I have no connection with this Pittman family for profiting in any way off the May tragedy. It was 4:42 p.m. in California and there is a one-hour time difference. He told me to call him December 5, 2014, the next day around 9:00 a.m., saying he was working with a family at that moment and he would provide the letter for me. He also gave me permission to use anything on the website below.

I further asked Mr. Pittman why the obituary had been shortened to just three sentences. He told me that Myron May's brother had contacted the chapel administrative office and requested the change. No one has found any information stating that Myron May had any siblings and it certainly was not noted on the program although he may have had. I suggested that perhaps it was the Uncle, who was spotlighted also in a televised news report saying the family was still in painful mourning. May's uncle said that when he last spoke with May, he was in a good mood and had planned to cook Thanksgiving dinner with the family. The family, who had known him since birth could not explain totally uncharacteristic violent behavior and were in complete and utter shock.

Mr. Pittman stated he could not remember the requester off hand and again asked me to call the next morning.

I then asked why his funeral home was chosen as opposed to the one in Wewahitchka. He replied that the Christian Memorial Chapel had been involved with several members of May family's passing of kin.

May himself confirms this in a post on his Facebook page August 2, 2014:

Myron May Facebook Post:

August 2, 2014

Uncle Cornelius, Aunt Gracie, Aunt Stella, and now Aunt Ada....my goodness Father. When it rains, it pours. Please slow down taking away my loved ones.

"Donald J. Pittman originally from Mariana, Florida, moved to Graceville, Florida along with his family fifty years ago and established a funeral home. The funeral home was called Pittman Funeral Home. He eventually changed the name to Christian Memorial Chapel the website states."

The Targeting of Myron May

Had not another Targeted Individual in the Facebook group TI's Stories captured the first obituary on Myron May, on November 26, 2014, listed on ChipleyBugle.com, there would not be a full version of the obituary detailing May's highly credible employment as shown below:

Within 24 hours, after publishing of the obituary above, detailing his accomplishments, ninety nine percent of the obituary had been reduced down to just three sentences as shown below.

Funeral services will be held Saturday November 29th at 11:00A.M. at Carter's Temple First Born Church of the Living God in Wewahitchka, Florida.

Interment will follow in the Buckhorn Cemetery in Wewahitchka under the directions of Christian Memorial Chapel of Graceville, Florida.

The question many would ask is why? Why was May's obituary which was initially published in its entirety, on November 25, 2014 deleted down to just three sentences on November 26, 2014? Why? One reason I believe is that the family did not want their names publicized in the media.

This blogger further stated:

It should be pointed out that all of the information relating to Renee Pittman Mitchell's connection to May, as well as that of Derrick Robinson, Robert Duncan, and Name Deleted, comes either from themselves or from non-mainstream news outlets, particularly the Memory Hole, none of which provide their sources for any of it. It is not too far of a stretch to assume that these outlets received this information from these people, as well as the altered suicide letters, and were willing to publish it without question. None of the mainstream news outlets that ran the May story mentioned anything about any of the people in the TI community who are now claiming to be an intended recipient of his packages. May's friend Joe Paul is the only person named by the authorities as being one of the intended recipients of the packages.

Was not this documented earlier, in three excerpts of interviews with an NBC (Universal) reporter, Reuters, and a televised news report by ABC Tampa Bay, Florida News 10. The Reuters reporter I actually gave a copy of the authentic suicide letter to written by Myron May and later substantiated as authentic.

On Saturday, November 22, 2014, I was contacted by a prominent individual within the targeted individual who is highly respected. I was asked if I would do an interview on "The Pete Santilli Show."

The Targeting of Myron May

Advertisement for this show was first in the form of a four-minute YouTube video revealing part of one of Myron May's voicemails shown at the following link. Creation of this advertisement was totally the idea of the show's producer after I agreed to the interview.

Typically, I am not active and never have been active in FFCHS, however, I learned while on the air that the Director was also a guest Below are links to my interviews and the advertisement the producer of this show created to announce that I would be on the show:

Pete Santilli Episode #848 – FSU Gunman Believed He Was Being "Targeted With Directed Energy"- deleted.

LINK: https://www.youtube.com/watch?v=E7HYM__Up_0

The actual three hour interview, Episode #848, is archived at the link below:

LINK: guerillamedianetwork.com/pete-santilli-show-episode-848-fsu-gunman-believed-he-was-being-targeted-with-directed-energy/

In a later blog image, this blogger writes of the factual altering of the Myron May suicide in which a person, I will use the initials, T. G. placed himself.

Note also that the mailing location he changed from, Tallahassee Florida, to Springfield, Ohio, which is documented as factually being a location of which Myron May had not lived until the age of 12 or 13. He moved to Florida during these years and the authentic documented postal tracking notifications substantiate the packages were factually mailed from Tallahassee, Florida.

Note also that the information at the top of the Priority Mail certified mailing package show's T.G.'s name. Note also these images are blurred due to the quality of the original images created by this person.

Finally, T.G's name has replaced mine as the official addressee and the courtesy copy information recipients have been adjusted to include the addition of three others.

It appears that this person thought that after the reports stated that all ten mailings had been intercepted and confiscated that this could be used to his advantage. As a result, he likely felt that tampering with the original letter could not be proven so he added himself boldly as a recipient. He later would tell others it is because those that May actually contacted did not care about May as if he did. Admittedly, I was not happy about what I felt was being used at all through my personal involvement and later what appeared to me to be isolated focus on me as a red flag.

In reality the effort of these two could appear to be typical tools of continued PSY OPS via disinformation, designed to confuse and distort information, and leave the impression that the letter and everything else was factually a hoax overall.

The authentic ten certified mailings were mailed from Tallahassee, Florida

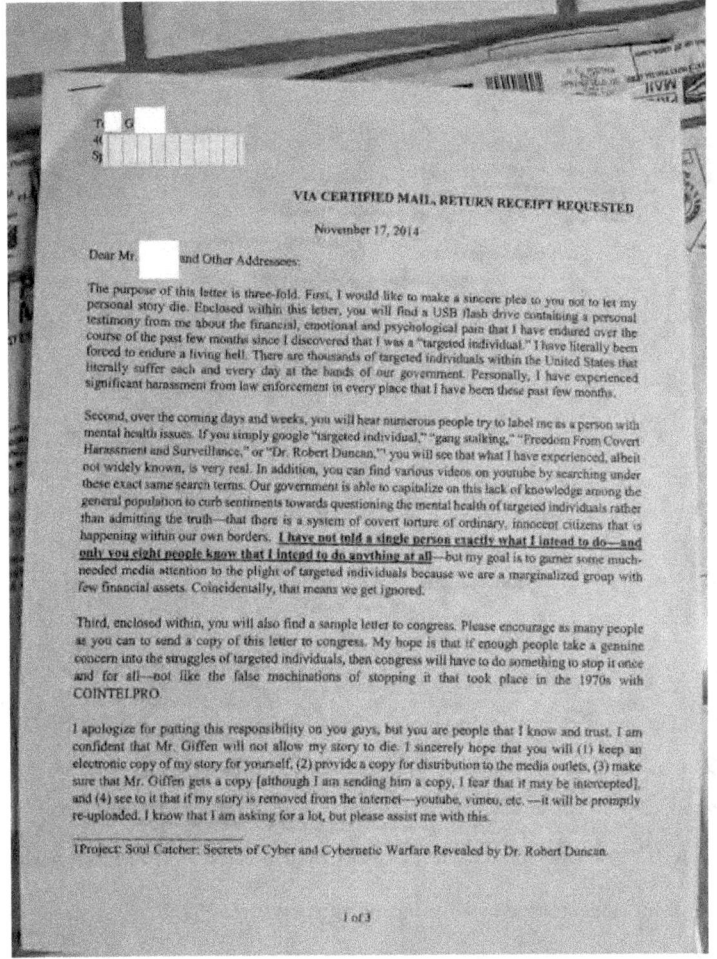

Altered letter where T.G. added his name "Mr."

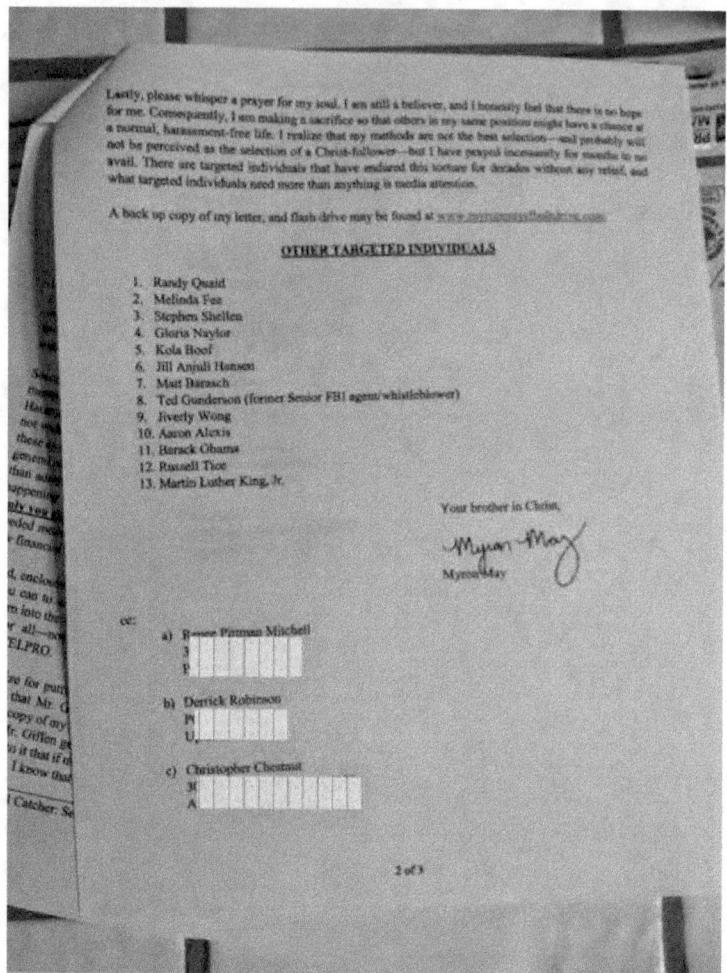

Three names are added, as number 11, 12 & 13:

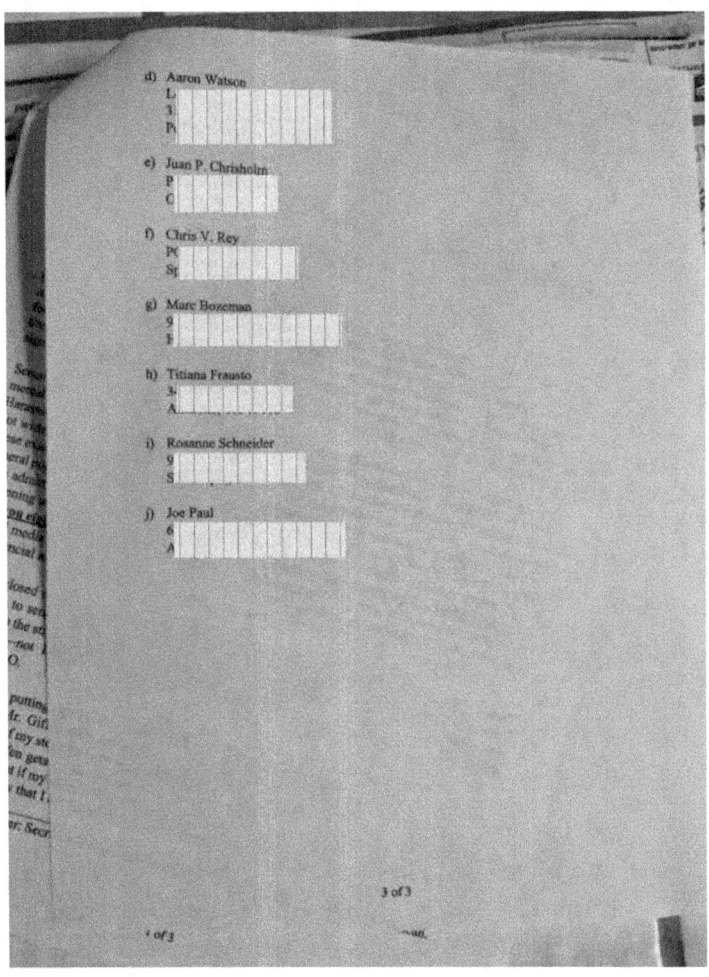

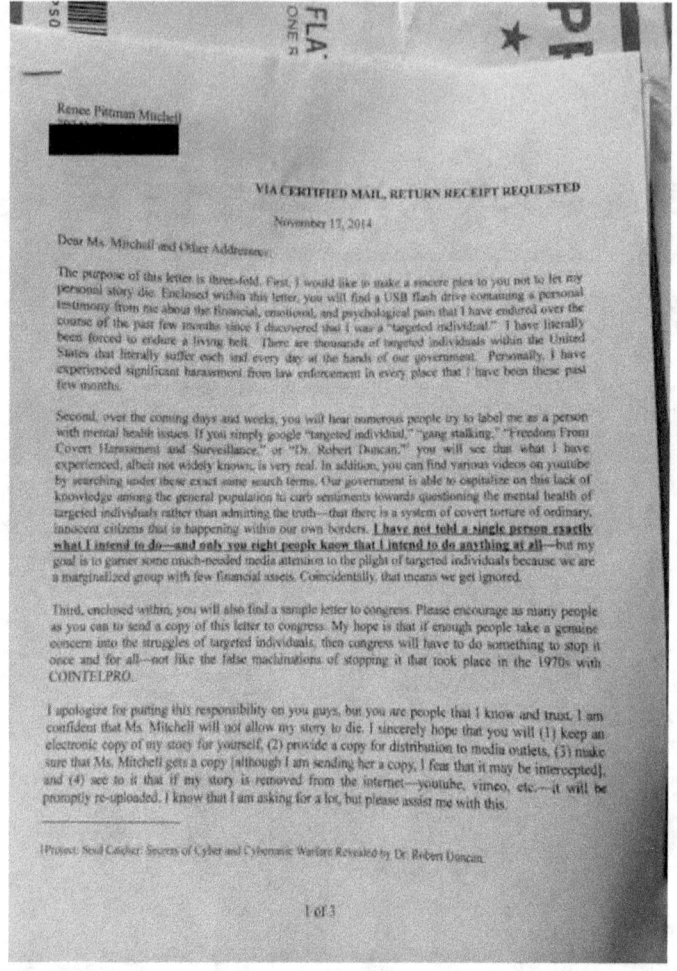

As revealed previously, this is the original letter emailed from Derrick Robinson, FFCHS, as shown previously. I apologize for the discoloration. It is due to the images being forwarded this way via email.

As far as I know, the videos were confiscated, destroyed, or deleted from YouTube and Vimeo if May ever had a chance to create and publish them. Again, all of the recipients report interception of the certified packages which included the flash drive disk. One thing is absolutely certain; T.G. was not listed on the authentic letter so he could not have a copy of anything.

What a bold farce T.G. embarked on after making questionable comments via social media condoning the actions that Myron May took and T.G.'s opinion that this is the only recourse for targeted individuals. His response and agreement of how Myron May handled his personal situation and how it took shape alarmed many.

While talking with someone on the phone and explaining the T.G. alterations of the letter and packaging, as an act of desperation to draw attention to T.G.'s personal plight, it appears, in the background of the conversation one of the African American supervisors could be heard saying, "Well we can't use him to get rid of her for us now." Believe it or not! T.G. appears to be heavily targeted he reports and he also admittedly reports also having spent time in the state mental hospital as a result.

Yes, the names of these PSY OPS games are intense creation of fear from these operations. Once they get the target with this, adding covert physical torture, no one believes is factually happening to them, and threats of imminent death, many become pliable.

Afterward, I started thinking about what the man said:

I thought Hollywood could not write a better script. Imagine T.G., who is on the record and who also has been proven to be extremely angry with me for not supplying him information and of whom now factually has my home address. Imagine operation center nudging to intensify his anger directed at me via radio frequency technology and being nudged to my home. Again, T.G. factually documents his hatefulness in vicious attacks on Facebook which someone sent to me and all targets are monitored by every means and around the clock. He was highly agitated after his failed attempts to get information from me to post onto his website. It is also documented on Facebook that he also called me several vindictive, vicious names as a result.

He literally also begged me for the emails, text messages saying that since posting information on Myron May on his websites, the traffic increased from a mere 200 to 700 views. That would mean a lot for someone trying to raise money for legal and medical costs as he states also on one of his sites.

T.G. would then show up at my door, and the next day headlines would read that yet another paranoid schizophrenic / so called targeted individual had kill someone this time Author Renee Pittman M., also a targeted individual. Get the picture? This is what the operation center was trying scare me with this possibility.

John Lennon is on the record referring to those unofficially running this planet as psychopaths touching a nerve. Lennon was also under surveillance around the clock as an enemy of the state, aka today, Domestic Terrorist by J. Edgar Hoover. He was factually funding individuals considered by the puppet master, the global Elite, to be a threat to global efforts who were protesting against war, racial division, civil rights, and for middle class Americans and gaining in numbers into the thousands.

Lennon was shot dead, December 8, 1980, by Mark Chapman. Chapman was reported to be a crazed fan however, Chapman had never possessed a Beatle or John Lennon solo record it was later revealed. Many believe that Chapman was factually a mind control victim nudged into murdering Lennon in a typical strategic effort because the secret government wanted him out of the way.

John Lennon earned the love and admiration of his generation by creating a huge body of work that inspired and led. ""All we are saying is give peace a chance." -- John Lennon… By October 1969, "Give Peace a Chance" had become a universal chant at anti-Vietnam War demonstrations by individuals with an ongoing war agenda.

Let me be clear, when it is my time to go, I will gladly pass to the other side but not a minute before God's appointed time for me.

The Targeting of Myron May

It would be funny if it were not completely possible and happening today, reported in heinous covert scenes by many, many accounts of many credible people to include once insiders who worked these operations. And, believe it or not, I am basing this scenario off what was factually said.

There is factually, documented, patented, unclassified, brilliant technological capable of accomplishing this and any other, effortlessly today. It is done by triggers, and by keeping the target in a hypnotic state, of which Myron May, may not have understood the capability or full power of technologically.

I document in "You Are Not My Big Brother" several incidents of the satellite radar laser beam unlocking my front door effortlessly in hope to terrorize me with the possibility that someone was going to come into my home while I slept and murder me. I also document setting off my smoke detectors while my house was cooled by the air conditioning in the summer's heat while living in Arizona in 2009 by the radar laser beam.

Whether fact or fiction, I decided to write about the comment here by individuals who just don't know how to shut up after now four books, and who refuse to leave and continue targeting me via extremely low frequencies. Fortunately, by many accounts, the books are credible in documenting what is happening today via these covert operation centers and this one makes five in the series. I will not live my life in fear which is what they want to control and that of many. The escalations also could be attributed to the fact that neither I nor anyone else was supposed to have the information which was in the Gmail to me, or was I?

However, no one can explain the capability of satellites today better than the excerpt from the article below of which I have used before:

"The Shocking Menace of Satellite Surveillance" by John Fleming in the excerpt from Part II below:

"There are various other satellite powers, such as manipulating electronic instruments and appliances like alarms, electronic watches and clocks, a television, radio, smoke detector and the electrical system of an automobile. For example, the digital alarm on a watch, tiny though it is, can be set off by a satellite from hundreds of miles up in space. And the light bulb of a lamp can be burned out with the burst of a laser from a satellite. In addition, street lights and porch lights can be turned on and off at will by someone at the controls of a satellite, the means being an electromagnetic beam which reverses the light's polarity. Or a lamp can be made to burn out in a burst of blue light when the switch is flicked. As with other satellite powers, it makes no difference if the light is under a roof or a ton of concrete--it can still be manipulated by a satellite laser. Types of satellite lasers include the free-electron laser, the x-ray laser, the neutral-particle-beam laser, the chemical-oxygen-iodine laser and the mid-infra-red advanced chemical laser…"

As for me, as I stated previously, I am so very grateful that I did not switch over to talk with Myron May when he began calling me hours before he made that fatal decision it appears. He obviously had 100% lost the spiritual warfare battle. And also, so very grateful that May wrote on November 17, 2014, in the real suicide letter addressed to myself and other recipients, making it clear by stating:

I do recollect that when I asked May when the targeting had started. He told me a year prior in 2013. He said that after getting pulled over by police officers numerous times, in Texas and in Las Cruces, Mexico, and a negative exchange due to his frustration, after one specific incident, two weeks later, he began to hear what sounded like people talking about him through the wall. Sadly, this is the exact complaint many reports experiencing, by technology no one want to admit or publicize factually exists, due to unscrupulous life destroying use, and the capability to literally tamper with a person's mind and body by electromagnetic weapons.

The discrediting starts by first the target calling police. The next step is going to a hospital to tell a doctor what his happening seeking help and support. From that point on, no matter whom you are, the official label is applied and documented in the target's health records. It then becomes literally open season on the target because the belief is the target is mentally ill. This leaves the targeted open to be picked prodded, tortured, technologically harassed, stalked, and used as a test subject for advanced virtual reality technology documented capable of illusion management.

Many targeted in this program, have long come to realize that those working this program are not above designating anyone a person of interest, or connecting them with drugs as an excuse, as easy prey due to perceptions.

Factually with little to no terrorist, available to justify these numerous counter terrorism division operating at the local police level today, many have found that they at times entertain themselves working in shifts by using the technology to simulate women sexually, for example, for their real time human life reality shows by men, again of all races, creeds and colors, working in shifts and legally employed and who appear empowered by the capability to destroy a human life. Some would argue for sport, well tucked away in these centers and from miles away.

In fact, I 100% actually heard the black men around me while I was on the phone talking with someone about the Myron May incident, in the background state that May became a target first out of sheer old fashion jealousy by white cops of a young black man earning more than they and having attained a level of prestige.

It is now my belief that after complete devastation, sacrificing another black man continues to be nothing new or a big deal. May learned quickly that no one has ever made it out of "The Program." This is again a program that involves all races that are horrifically being covertly victimized. This includes those he listed as prominent targeted individuals in his letters. He also must have grasped that many live with the ongoing torture, and harassment, and degradation, relentlessly, around the clock for many, many years and lives never returning to normal. And, by numerous reports, many new targets are being added each and every day.

May actually told me that he resigned from his job at the District Attorney's office because the technological verbal harassment was unbearable and that as a result he could not function any longer to do his job properly which was the intent.

Intentional sleep deprivation also played a significant and vital role in his demise both emotionally, physically, and mentally. This definitely also contributes to loss of will power and the reason it is part of the PSY OPS program. People become vulnerable with lack of sleep, and can make wrong decisions through deprivation. The battle against this powerful technological Goliath is not for the faint of heart of both men and women. Men can be deeply hurt too and disappointed, but many continue to fight the battle in a positive way.

Many of us have been fighting for our lives honorably as noted by Myron May when he posted an image of Jesse Ventura speaking about mind control online. Robert Duncan is an ex-insider who reported he was involved in the evolution of this technology for mass population control in the United States and was spotlighted on Jesse Ventura's Brain Invader episode. Many targeted individuals watched the December 2012 "Brain Invader" episode on TruTV, Conspiracy Theory and hoped that now mainstream media would begin to report the reality and use of the technology. It just did not happen. Targeted individuals are up against a decisive global effort where total population control and also gun control laws are pivotal for global success.

The fact is those spearheading these heinous programs are just as guilty and more so if they were technologically behind May's snapping.

Early morning, Thursday, November 20, 2014, I woke around 7:30 a.m. with Myron May on my mind. In spite of my uneasiness about him, I decided to text him.

I reached for my cell phone and listened to the voicemails. I then immediately texted the following to the woman in Northern California that someone needed to reach out to him but it would not be me. She was as I mentioned at work:

Renee Pittman M. to associate in Northern California via FB

Thursday, November 20, 2014, 10:05am

Hi. I finally checked the voice messages Myron left last night while on the phone with you. He said that they are hitting his heart repeatedly with the beam and that he does not have long to live. He said that he devised a plan. As I mentioned last night, it appears that he believes that a kamikaze move will relieve him of the torture and bring light to our plight. He sent me two receipts for regular stamp purchases only totaling $24. He begged with me to call him back and not leave him during this time and gave the impression I am part of the plan.

The problem with me is that I don't believe him 100%. Most people don't want to kill themselves and if I do, I can do nothing to stop him in Florida. I tried to dissuade help and hope to give him an understanding of how this program works. I explained how the program is structured and that a primary tactic to control people is by beaming suicide thoughts and by fear of death. I explained that we all have experienced this and that many of us are still standing because we do not fear death. In my personal opinion death is a return home and I will gladly return.

To me, this sounds like a plea for help with the exception of likely Operation Center pulling his strings. I told him that exposure is pivotal in these efforts through activism and even offered to renew the Blog Talk Radio show with him as host.

I also hoped he could become pro-active by helping both you and me legally. When we got off the phone he seemed motivated. That was until I noticed that it is likely he had not even mailed anything certified. Showing purchases of regular stamps means nothing when typically, confirmation of mailing is identified by the Post Office by stamping the green and white receipt.

I texted an example to you. I don't know what to do or how to help him. Do you have any suggestions? Would you like to call him? I don't want to be involved if he is really determined to commit suicide. That is a choice between him and his God. I had never seen, nor heard of him, as you stated also until 5 days ago.

CORRECTION: I tried to dissuade "him" not dissuade help…

CORRECTION: Most people do not want to kill themselves and if "they" do… Not - "if I do" …

I personally am in it until God calls me home. I also, as I mentioned, in the text I sent you after we got off the phone am offended that I offered to help him, help himself and others through use of his law degree and he brushed me off yet still offered his services elsewhere. Nothing personal with you, but this is another red flag with me which I do not understand.

Do you want to call him? I can't I have tried to talk to time twice in lengthy phone conversations. Plus, I can be blunt/direct without malice, and don't want to do or say anything to evolve this situation.

I have always been this way.

I just text the following message to him before I head out the door:

"If you kill yourself you hurt our effort and allow this program to continue. You can bet the media will portray before millions as Schizophrenic. They will even use you ex Pediatrician doctor girlfriend to substantiate that you told her you were hearing voices and others.

Don't involve me.

Psychotronically aka Psychological Electronic

FACEBOOK Chat Conversation End

To my dismay, Myron had already fulfilled his scheme, little known to me as I wrote this Facebook message.

Many of us have felt like we could not go on at one time or another, and have endured, and continue to endure. The victory is so close. With so many voices saying the same thing eventual disclosure is on the horizon. Let's hope it does not take 20 years and the Church Committee hearings as it had for the government backed MK ULTRA mind control program.

Yes, many of us did not lose a high paying job in which racial hatred and racial profiling appears to have played a role; by Myron May's own admission of his account on Facebook but have suffered great losses also.

Intuitively I tried to back off from him several times as I watched his focus zero in on me.

After the fact, when I read back over the texts, email, and voicemails, it was clear that an impression could have been given by Myron May, to an outsider that we knew each other personally or perhaps had known each other much longer than six days. I had to ask myself could this have been intentional. Any intelligent person could see that I had not caused what happened. And as I said later, in the ABC televised news report, had I known what his plan was, I would have surely called police in Florida and hopefully saved his life and prevent injury of the others. Again, I thank God my instincts, although unsure, cautioned me to try to back away from him several times.

I was not trying to hurt him, influence, or manipulate him in any way. Let it be known that I am not the go to girl as the battle rages for my own life. This is as many began contacting me asking could I help them revealing to me an ongoing PSY OPS around me even after this tragedy. Let it also be known that my life is nothing to play with however those targeting me continue to believe my life and so many other lives are on so many levels and for various strategically useful reasons.

Regarding ongoing racial harassment by profiling May, in August of 2014 posted his experiences on Facebook related to ongoing police harassment via traffic stops. These efforts, too as a target, can be designed to push a person over the edge as overt harassment. Today if combined with a technological capability, around the clock it can be devastating.

The Targeting of Myron May

Myron May posted the following image and story on Facebook:

After A Traffic Stop, Teen Was 'Almost Another Dead Black Male'

Patsy Hathaway, who is white, thought "love would conquer all" when it came to how others would treat her adopted son, Alex Landau. That changed after he was severely...

npr.org | By Morning Edition

Myron May Facebook Post Response - August 19, 2014

Speaking from personally having experienced about 10 DWBs (Driving While Black), YOU DON'T GET MEDIA COVERAGE FOR IMPROPER POLICE CONDUCT UNLESS YOU DIE. So here are.

The Three that Made Me Most Angry:

#1**Hats off to my buddy NAME DELETED who appeared out of nowhere when an officer was about to arrest me in Tallahassee b/c according to him, he ran my license plate, and his driver's license check indicated that my DL was suspended (it was not). In light of the fact that his DL check also indicated that I was a white male, you'd think that he would have questioned the accuracy of his info. Nope!

#2 I got pulled over in my own neighborhood of Katy, TX a few times too. But in one instance, the cop just told me straight up that he thought I looked "suspicious." I was driving a new Equinox (soccer mom car). I'm not so sure what's so suspicious about a soccer mom car in suburbia, but go ahead and run me through prior's sir so that I can go to my house and sleep.

#3 when I was in law school in Lubbock, I made a left on campus where the speed limit was 20mph. There was an officer coming in the opposite direction. As we passed, he immediately makes a U-turn and stops my car. At this point, I've made it at most about 200 yards from the traffic light. He comes to the window and says, "I paced you going 30mph." Sir, am I a fool? First, you can't pace me traveling in the opposite direction as me. Second, it isn't mathematically possible for me to have traveled at 30mph and only be this far from the traffic light. Name Deleted kept me calm on that one.

**Thankfully, I've never been beaten or killed. But to my friends that think this sort of thing doesn't happen in today's world, please don't be that naive. It happens every day. I wish I could explain the degree of rage, frustration, resentment, and bitterness that you feel after having one of these experiences, but I don't know that ordinary language is even comprehensive enough to convey that idea. So, you have to choose to either let what little bit of mental fortitude that remains serve as an anesthetic for those emotions, or you're going to become so inundated with them as the encounters increase that you are literally a ticking time bomb. For me, it was like a pendulum between anesthetic and time bomb. But now, having worked as a prosecutor, I know that only a small fraction of cops does this sort of thing. The problem is that they do it with such frequency that their impact on citizen-cop relations is expansive, which coincidentally impedes officers that are just trying to do their job in a legitimate way.

As I stated which can be verified via my website bigbrotherwatchingus.com at the Technology Approval tab, this technology has trickled down from the DOD to the DOJ and legalized after 9/11 under the guise of the "War on Terrorism" which combined with DHS efforts results in anyone being targeted for any reason under the Sun who can be placed on a Watch List for people looking for something to do during each 8 hour shift.

For example, the LAPD's Real-time Analysis Critical Response (RACR) Division is an example of satellite and drone capabilities at the local police level on steroids. Black men today are factually under the gun make no mistake about this. This is especially true of those working these operations trying to earn a living anyway they can as all others in these operations. And, due to the ability to hide, and use technology, as I have stated before technologically designed as if playing a video game will kill in a country effectively desensitized to murder for decades by subtle programming.

For example, Ferguson, Missouri, Michael Brown wasn't gunned down as a fleeing felon or criminal and there are overtures of racism. The fact is Missouri is a former "slavery" state once openly and violently controlled by the Confederate General Albert Pike's (I reckon with the Dragon) Illuminati Knights of the Golden Dawn.

Michael Brown was slaughtered for the unwritten "Code of Insolence". Like Trayvon Martin, Jordan Davis and so many young black males and technologically it appears Myron May. If a white man challenges you, you are not supposed to move a muscle to defend your rights of yourself. The public is groomed to believe that in the instance of their deaths that young black men, no matter how successful, were suddenly transformed into uncontrollable and raging maniacs that threw themselves into a hail of bullets. If May had not shot first, could he have been taken down, non-violently, and prosecuted? Sadly, not relinquishing his weapon when asked played a role either way.

While I totally understand this, some would argue that I go after black men in my books depicting them in a negative light thereby contributing to a honed perception of black men designed to make extermination of black men easier. I owe no allegiance to any specific race especially those around me causing me great pain and serious health problems, and who have attempted to effectively destroy me and my family. I stand aligned with the human race and with the good people of all races both black and white and other as all should for what is right and against what is wrong. I simply cannot protect or condone the work of the specific group around me, whites included in supervisory positions.

In my case, because I have taken to charting the slow deterioration via MRI and x-rays to reveal unusual microwave directed energy weapon slow creation of illness and deterioration to my body and the biological effects. They would like to kill me outright in the effort to silence exposure. I seem to be a challenge although several people have written similar books on this subject. Fighting back makes you interesting to their boring psychopathic lives in these centers and the urge for desired using anyone for entertainment are welcomed challenges to extreme levels.

The reality is that books afford the opportunity to take the message of what is happening today globally, even if it is one book at a time in the human chain. Book publication can effectively challenge a strategic programmed perception of what is real and what isn't based on factual credible evidence. They can also stand as an equalizer and vehicle of defense in a climate where mainstream media is not reporting fact-based reality today. No one in any news report mentioned that factually directed energy weapons are very real and in full use everywhere today. They focused on a man, deemed mentally disturbed, originating out of the blue, and it appears for no reason whatsoever went Postal, who prior to a few months before had years of credibility.

The question is who is really mentally disturbed, albeit legitimately as revealed in except below:

US Police Have Killed Over 5,000 Civilians Since 9/11

Statistically speaking, Americans should be more fearful of the local cops than "terrorists." By Katie Rucke

November 10, 2013 "Information Clearing House - "Mint Press News" --

Though Americans commonly believe law enforcement's role in society is to protect them and ensure peace and stability within the community, the sad reality is that police departments are often more focused on enforcing laws, making arrests and issuing citations. As a result of this as well as an increase in militarized policing techniques, Americans are eight times more likely to be killed by a police officer than by a terrorist, estimates a Washington's Blog report based on official statistical data.

Though the U.S. government does not have database collecting information about the total number of police involved shootings each year, it's estimated that between 500 and 1,000 Americans are killed by police officers each year. Since 9/11, about 5,000 Americans have been killed by U.S. police officers, which is almost equivalent to the number of U.S. soldiers who have been killed in the line of duty in Iraq…, she writes.

And, I might add, society should be even more fearful of the technology in the hands of these programmed minions of the NWO disguised through mass perception as honorable patriots to America and U.S. citizens. The death toll continues to rise at the hands of the now technologically mobilized forces with their new covert, deadly toys, in the hands of some who come to the table with extreme, dysfunction mental issues and with something to prove to the world now elevated to authority by a higher echelon with the power to kill legally and get away with it we continue to see.

Laws of ethical, humane treatment of everyone are in place for a vital reason. Without which there would be anarchy which is the case today, covertly. Make no mistake about it racism thrives in law enforcement. A recent USA Today article reported that a Florida Police Chief and deputy both resigned after the FBI revealed that both belonged to the Ku Klux Klan. The reality is that actually law enforcement; at ALL levels has a deep history with racial identity materializing after the Civil War as efforts first to keep the freed slaves in place through horrific inhumane violence and murder and programming that "Black lives do not matter."

Although, today, many reports that military operations are factually doing some of the most heinous and advanced technological testing and manipulation of everyone. These testing programs are happening within the USA also in NATO countries. The new paradigm is Fusion.

One must ask who is factually benefiting from mass population control. If this can be answered with focus on one specific agency, it would be easy. However, there are multiple players. The answer would reveal the puppet master pulling the strings and ongoing efforts of devaluing of human lives across the board making anyone a patsy for globalization and depopulation of which we are witnessing right before our very eyes.

CHAPTER SIX
Powerful Technology at Play Today

After the Myron May incident and as I begin this project, three drones that had been positioned over my home each night beaming the Sonic Weapon this was at first typically while I tried to sleep at night, attacks were immediately intensified. I also noticed that neighbors were being incited yet again to mobilize stalking efforts. While I worked one of the supervisors stood outside of my door at night saying, "She's gone" then set up shop in a neighbor's home where they could beam vicious names into my home. In reality with this supervisor showing up in my neighborhood, they were setting the stage to hope to terrorize me by entering my home aka gas lighting. This is what the tapping noise on my front door was about the night before. As I wrote this book, they intended to leave the message that if I continued they would enter my home and murder me.

Names mean nothing unless you believe they are true and you suffer from low self-esteem. It appeared that the operation center employees were again pulling out all the stops, and now hoped to step up the fear campaign. As stated, when writing the other books, I had lived through portable microwave beams deployed from every single house positioned nearest to mine front, back side by side and a few at various strategic angles to rooms in my house such as my bedroom which I mainly stayed working at my desk.

I live about 20 miles from a major USAF base and one day unbelievably a black helicopter circled my house as I worked telling a tell-tale sign of United States Air Force involvement in these operations also and beamed my window and the window rattled. The amped frequency, of the sonic energy weapon was no longer targeting me exclusively during the night but now all day long. Without a doubt, the intent was to escalate neurological damage, brain cancer or aneurysm.

The greatest characteristic and asset of these electromagnetic weapons are there capability to kill without leaving a trace or revealing that energy weapons are the culprit. There affect is undetectable. If a person is hit in spurts with this bio-coded weapon system then it is survivable, as a non-lethal attack, that is unless consistent attacks. Escalated covert murder was on the agenda for me after the May incident appeared to have failed to pull me into "a scheme" possibly willingly or unwilling. Again, based on his reporting that no one knew of his plans, it had factually saved my life.

It is not out of the question to consider that Myron May could have been held in a virtual hypnotic trance via powerful technology either. I began to believe this as a possibility as the dust settled and my familiarity of this program through research reveals it as possible.

The Directed Energy Weapon system, which is deployed in many formats, is a bio-coded system using Biometric Authentication. The fact is Directed Energy Weapons (DEW) of which May, repeated complained of being hit, relentlessly, are playing key roles in FBI/military PSY OPS, and many targeted individuals, report also attacking of key areas of the target's body which includes the chest area. Understand that these weapons, although documented as "non-lethal, a misnomer, can be lethal if use to kill and can do so immediately if so desired.

Without a doubt, in Myron May's case, any hopeful unified effort of attorneys could become a great threat via law degrees and as worthy adversaries to expose this program. Most, if not all Plaintiff's in these cases are forced to go it alone, Pro Se. As a result, there is little hope for a victory when up against high powered government U.S. Attorneys.

I believe that with Myron May, there might have been hope if he had only decided to stand and fight. The reality is that it is very difficult to get attorneys on board legally in these cases. I have been personally told that they know what is happening today, as I have said before, but are afraid that their lives will be ruined just as May's had.

The term "bar" is often used to mean "the legal profession". Thus the "bar exam" is a test taken to determine if one is qualified to be licensed as a lawyer or "member of the bar".

The American BAR Association (and its State alter egos) has, for all intent and purposes, taken over our entire federal, state, and local governments. The legislative branch follows the advice of their BAR member advisors in constructing of statues. The judicial branch is literally a closed union shop in that regard. You cannot be a judge unless you are a BAR member and you cannot practice in their courts unless you are a BAR member.

The term "BAR" is an acronym for British Accredited Registry. In fact, this means that judges and attorneys are in fact working for the crown of England. This is why gold fringed flags are in courtrooms, and other official locations. It signifies admiralty jurisdiction (maritime law), which is another way of saying British jurisdiction. England is a maritime nation. When you cross the bar in a courtroom, you are entering a British colonial forum.

The NWO began when this country was founded

Gold fringe on flags behind President Obama

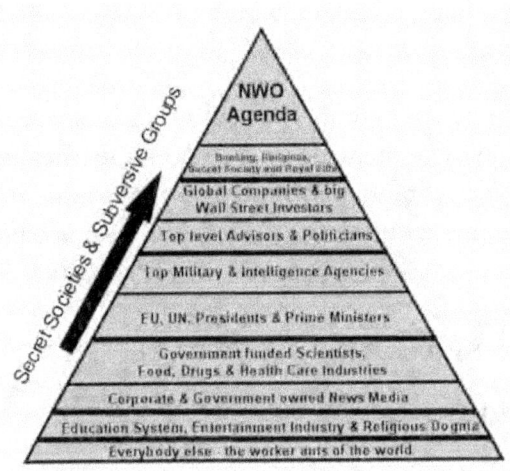

Those of us who continue to expose what is happening, in many forums including hopeful court intervention, have been left out in the cold through an inability to effectively try these types of case as Pro Se, as our own attorney.

Documented in Congressman Dennis Kucinich's HR2977 also known as "The Space Preservations Act of 2001, and also retired Missouri State Representative James Guest, October 2007" letter to his constituents, are obvious hopes to expose and curtail what retired State Rep Guest calls "Electronic Weapons Torture, on United States citizens to include women and children, in and of itself should reveal the reality if not the capability of Directed Energy Weapons

I again offer for view: Dennis Kucinich's Space Preservation Act of 2001 below:

HR 2977 IH

107th CONGRESS

1st Session

H. R. 2977

To preserve the cooperative, peaceful uses of space for the benefit of all humankind by permanently prohibiting the basing of weapons in space by the United States, and to require the President to take action to adopt and implement a world treaty banning space-based weapons.

IN THE HOUSE OF REPRESENTATIVES

October 2, 2001

Mr. KUCINICH introduced the following bill; which was referred to the Committee on Science, and in addition to the Committees on Armed Services, and International Relations, for a period to be subsequently determined by the Speaker, in each case for consideration of such provisions as fall within the jurisdiction of the committee concerned

A BILL

To preserve the cooperative, peaceful uses of space for the benefit of all humankind by permanently prohibiting the basing of weapons in space by the United States, and to require the President to act to adopt and implement a world treaty banning space-based weapons.

Be it enacted by the Senate and House of Representatives of the United States of America in Congress assembled,

SECTION 1. SHORT TITLE.

This Act may be cited as the `Space Preservation Act of 2001'.

SEC. 2. REAFFIRMATION OF POLICY ON THE PRESERVATION OF PEACE IN SPACE.

Congress reaffirms the policy expressed in section 102(a) of the National Aeronautics and Space Act of 1958 (42 U.S.C. 2451(a)), stating that it 'is the policy of the United States that activities in space should be devoted to peaceful purposes for the benefit of all mankind.'.

SEC. 3. PERMANENT BAN ON BASING OF WEAPONS IN SPACE.

The President shall--

(1) implement a permanent ban on space-based weapons of the United States and remove from space any existing space-based weapons of the United States; and

(2) immediately order the permanent termination of research and development, testing, manufacturing, production, and deployment of all space-based weapons of the United States and their components.

SEC. 4. WORLD AGREEMENT BANNING SPACE-BASED WEAPONS.

The President shall direct the United States representatives to the United Nations and other international organizations to immediately work toward negotiating, adopting, and implementing a world agreement banning space-based weapons.

SEC. 5. REPORT.

The President shall submit to Congress not later than 90 days after the date of the enactment of this Act, and every 90 days thereafter, a report on--

(1) the implementation of the permanent ban on space-based weapons required by section 3; and

(2) progress toward negotiating, adopting, and implementing the agreement described in section 4.

SEC. 6. NON-SPACE-BASED WEAPONS ACTIVITIES.

Nothing in this Act may be construed as prohibiting the use of funds for--

(1) space exploration;

(2) space research and development;

(3) testing, manufacturing, or production that is not related to space-based weapons or systems; or

(4) civil, commercial, or defense activities (including communications, navigation, surveillance, reconnaissance, early warning, or remote sensing) that are not related to space-based weapons or systems.

SEC. 7. DEFINITIONS.

In this Act:

(1) The term `space' means all space extending upward from an altitude greater than 60 kilometers above the surface of the earth and any celestial body in such space.

(2) (A) The terms `weapon' and `weapons system' mean a device capable of any of the following:

  (i) Damaging or destroying an object (whether in outer space, in the atmosphere, or on earth) by--

    (I) firing one or more projectiles to collide with that object;

(II) detonating one or more explosive devices in close proximity to that object;

(III) directing a source of energy (including molecular or atomic energy, subatomic particle beams, electromagnetic radiation, plasma, or extremely low frequency (ELF) or ultra-low frequency (ULF) energy radiation) against that object; or

(IV) any other unacknowledged or as yet undeveloped means.

(ii) Inflicting death or injury on, or damaging or destroying, a person (or the biological life, bodily health, mental health, or physical and economic well-being of a person)--

(I) through the use of any of the means described in clause (i) or subparagraph (B);

(II) through the use of land-based, sea-based, or space-based systems using radiation, electromagnetic, psychotronic, sonic, laser, or other energies directed at individual persons or targeted populations for the purpose of information war, mood management, or mind control of such persons or populations; or

(III) by expelling chemical or biological agents in the vicinity of a person.

(B) Such terms include exotic weapons systems such as--

(i) electronic, psychotronic, or information weapons;

(ii) chemtrails;

(iii) high altitude ultra-low frequency weapons systems;

(iv) plasma, electromagnetic, sonic, or ultrasonic weapons;

(v) laser weapons systems;

(vi) strategic, theater, tactical, or extraterrestrial weapons; and

(vii) chemical, biological, environmental, climate, or tectonic weapons.

(C) The term `exotic weapons systems' includes weapons designed to damage space or natural ecosystems (such as the ionosphere and upper atmosphere) or climate, weather, and tectonic systems with the purpose of inducing damage or destruction upon a target population or region on earth or in space.

Myron May including Congressman Kucinich as a person to contact on his second Targeted Individual "To Do" list is likely for this reason.

State Representative Jim Guest Letter Written October 2007:

The Targeting of Myron May

October 10, 2007
Dear Member of the Legislature and Friends:

This letter is to ask for your help for the many constituents in our country who are being affected unjustly by electronic weapons torture and covert harassment groups. Serious privacy rights violations and physical injuries have been caused by the activities of these groups and their use of so-called non-lethal weapons on men, women, and even children.

I am asking you to play a role in helping these victims and also stopping the massive movement in the use of Verichip and RFID technologies in tracking Americans. Long before Verichip was known, we were testing these devices on Americans, many without their knowledge or consent. With the new revelations of cancer risk beside the privacy and human rights problems with the use of Verichips and RF signals, I am asking for your assistance in stopping these abuses and aiding those already affected. Your attendance is therefore requested at a conference call regarding these issues on Monday, October 29 at 11 am, EST. After a period of brief presentations, we will have a discussion of these issues with the intent of creating a way forward solutions...

Sincerely,
Representative Jim Guest

Noteworthy, in May's case, within a year of his initial targeting, he was escalated to one of the highest levels in this program, through possible Delta Programming, which is documented to be part of MK ULTRA Manchurian programming experiments, and marked; it appears for termination possibly for a cause. Many, who realized that something is up, as they watch their lives being destroyed in a very surreal fashion electromagnetically while monitored around the clock, when factually connecting the dots, are taken then to the energy weapon torture level. May being very bright caught on fast. And, possibly became a threat and became usefully expendable.

Since 2005, Hewlett Packard (HP) has been on the open market selling countries national identity systems built on the .Net platform.

Hewlett Packard and Microsoft have worked on a code base that will allow them to offer a set of technology components for functions such as online and offline demographic and biometric data capture, regional verification and registration, and document lifecycle management. Biometrics (or biometric authentication) refers to the identification of humans by their characteristics or traits i.e., DNA, iris, gait, voice and facial recognition, biometrics is used in computer science as a form of identification and access control. It is also used to identify individuals in groups that are under surveillance.

For example, since 2006, Hewlett Packard has been also investing heavily in the buildup of the software/computer surveillance technology- check point (biometric ID system) and telecommunication capabilities.

It is a daily practice today in warring nations, and globally of a necessity for technological surveillance and control. In recent years, these practices have increasingly relied on technological mechanisms provided by international and local corporations. Hewlett-Packard is one of the companies that enable this technological supervision and oppression.

Hewlett Packard is the prime contractor of the Basel System, an automated biometric access control system installed and maintained by Hewlett Packard checkpoints globally. Hewlett Packard is involved, in also an ID card system, which reflects and reinforces the global state's political and economic asymmetries as well as its tiered citizenship structure through use of biometric ID cards.

Two of Hewlett Packard's technological services provided Matrix and its subsidiary, Tact Testware and further participates in the "Smart City" which provides a storage system for the settlement's municipality in other countries.

"Among their array of evermore advanced intelligence tools, the Pentagon and NSA have cooperated on a Hewlett-Packard project called "The Swarm Enterprise."

"The Swarm is essentially the precognition program depicted in the Tom Cruise movie [Minority Report] on pre-crime prevention." The Department of Homeland Security system in use from Fusion Centers within United States as stated previously Malintent.

The technology is not without controversy. An internal U.S. Department of Homeland Security document indicates that a controversial program designed to predict whether a person will commit a crime has already tested on some members of the public involuntarily, CNET has learned. If this sounds a bit like the Tom Cruise movie called "Minority Report", or the CBS drama "Person of Interest," it is.

But where "Minority Report" author Philip K. Dick enlisted psychics to predict crimes, DHS is betting on algorithms: it's building a "prototype screening facility" that it hopes will use factors such as ethnicity, gender, breathing, and heart rate to "detect cues indicative of mal-intent."

What is not mentioned is that this technology is also mind reading or thought deciphering software using advanced EEGs. This is where the mimicry comes in by those in the operation center sitting and repeated the target's thoughts back to them which ultimately is subtle terrorism.

Excerpt from internal DHS document obtained by the Electronic Privacy Information Center at the link below:

See more at:

http://www.darkgovernment.com/news/controversial-dhs-malintent-detection-technology/#sthash.lrHhYfiT.dpuf

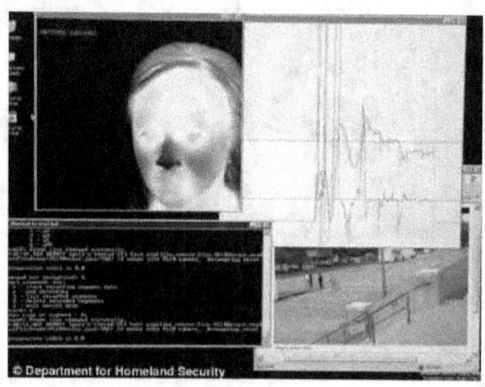

The Targeting of Myron May

Within the Enterprise Swarm Research Program Hewlett Packard is involved in developing a computer software system that develops the state's criminalization of notions of "precrime" and "thought crimes."

It would develop models to project future crimes then police or the military could arrest the potential criminal and stop the crime before it happens or target the individual then marked for human experimentation.

This is similar to what George Orwell, in his novel "1984" refers to as "Orwellian Thought Crimes and today, literally the "Thought Police" are in full force as part of the operations in these centers. For example, if a person is having a bad day, and even thinks of committing a crime, as even a passing thought, a person is subject to be arrested by the state under Swarm conjures.

The fact is we have reached the age of precrime monitoring where people become suspects even if they have not done anything wrong. The most common excuse by society today to allow it without protest is "I have done nothing wrong so I have nothing to hide." The problem is that people do not know what is actually being used against them. Even today, social media posts can impact your threat rating. These are not Homeland Security threat ratings but ratings by your local police department which, again, are soon to be federalized. U.S. citizens are factually getting a colored coded threat rating every time you encounter police who are using numerous tools and applications. One such application is "Beware" sold to police departments in 2012. The "Beware" algorithm assigns a score and "threat rating" to a person of green, yellow, or red and sends the information to a requesting officer. How and why your rating materialized is not made available to you. For example, if you wrote something about recent Ferguson protests on Facebook. Perhaps you expressed dissatisfaction with police brutality. Maybe you complained about the torture tactics or demanded Obama's impeachment or criticized the government in general. Any comment which can be contributed as offensive can contribute to your threat score and lead to amped harassment.

What if May, or anyone else, in frustration said he hated law enforcement, after ongoing, racial profiling, terrorism and extreme technological harassment. In most cases, human reactions simply pass and the anger subsides. Logically, this would be a human reaction of which he never would have acted on negatively. However, what if he was then placed into "The Program" and those monitoring him deciding they don't like him, or even deeming him an Uppity Negro, his life then becomes their toy?

Many believe that HP's Swarm Project is just another veiled backdoor and repackaging of Dr. Hutshnecker's SS Plan for the justification of the state to demonize, criminalize, harass and place people of color in concentration camps for or today what are technological concentration camps determined to have predisposed "anti-social tendencies" using race algorithms to nudge anyone, specifically certain despised races, into criminal activity, then create a media perception of the manipulated and influenced mass controlled individual, groups, etc. as deserving of genocide.

Many of these ideals have for decades, believe it or not, focused on African Americans and historically, the African American community have always been the first stop and testing ground for mind control, manipulation and influence technology, today using Biometric Signatures and radio frequency technology which is set to an algorithm which can be race specific. The software can then trigger and ignite African Americans into agitation, anger, panic, etc., which Ferguson could be used as a possible prime example in a goal to usher in Martial Law.

These practices have a doctrine in ongoing mind control studies originating in Nazi Concentration Camps and afterward continued underground studies and efforts connecting Operation Paperclip when Nazi scientist who were given refuge in the United States to continue mind control scientific research. At the foundation, these concepts are based on eugenics, Madame Blavatsky's Luciferian Root Races, Secret Doctrine, and Nazi pseudo race theories and what was gained through decades of mind control testing efforts which lasted for over 20 years officially within the United States in again, the supreme mind control program within America of which the current program has its foundation called MK ULTRA.

In many ways, America had been the mentor of the Nazis. But, the SS and Nazis introduced into America a new level of ultra-occulted secrecy, state oppression and fascism, and covert campaign of mass racial genocide deeply imbedded by its blood money and satanic blood rites. Believe it or not! The wrong is in denying the value of all life and its negative effort on humanity as a whole.

Tens of thousands of the "Lucifer's Servants" most notorious were clandestinely implanted in key positions of influence, authority and power throughout America's major corporations, mass media groups, universities, hospitals, foundations, governmental agencies, military branches and the Pentagon; the FBI and the Central Intelligence Agency (CIA) to include State and local law enforcement and the military.

They created the CIA and America's National Security State, the National Security Council (NSC), and the National Security Agency (NSA).

The creation of NSA resulted from a December 10, 1951, memo sent by CIA Director Walter Bedell Smith to the National Security Council to create a twisted sister agency to compile and collect centralized intelligence data on (all) U.S. citizens.

The NSA was formally established through a revision of National Security Council Intelligence Directive (NSCID) 9 on October 24, 1952. It is probably the most secretive branch of the Invisible Government. Even more than the CIA, the NSA has sought to conceal the nature of its activities but you can bet is playing a powerful role in the globalization effort. The only official description of its activities is contained in the U.S. Government Organization Manual, which states vaguely:

"The National Security Agency performs highly specialized technical and coordinating functions relating to the national security."

The fact is, new mind control technology has been ongoing for decades and has resulted in highly advanced technology.

The website techswarm.com reports, November 24, 2014, how "New Research Shows How Virtual Reality Technology Shuts Down the Brain." Within the article it details how a person can be programmed and an excerpt reads:

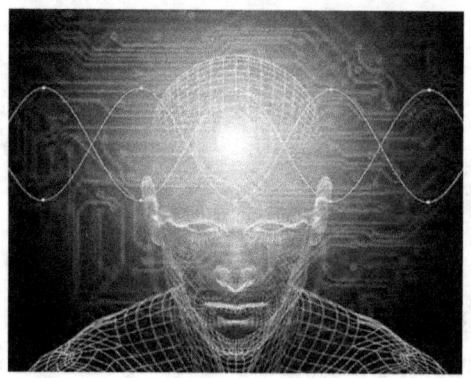

Excerpt from Nicholas West

Virtual reality is already being embraced for its entertainment value, as well as by the military and the scientific establishment. It is also a goal of The Singularity Movement to enable a full mind upload as we increase our merger with machines toward a path of supposed immortality.

However, scientists are beginning to study the effects of how virtual reality can impact one's perception of themselves inside the virtual matrix, as well as a potential for transferred perceptions of those around them in the real world.

Early conclusions are troubling. It appears that not only can our moral behavior be affected, but parts of our brain that register spatial awareness and movement actually shut down when entering even the most realistic virtual environment…"

What if mass shooters, minds are shut down by powerful biometric computers?

Related to deployment of Directed Energy Weapon attacks, the CIA spent three years training Colombian close air support teams on using lasers to clandestinely guide pilots and laser-guided drones to their targets by special GPS software tracing biometric authentication, computers, hard drives and cell phones calls.

In the United States controversial drone strikes involving the biometric authentication" targeted killing" has not been fully revealed to the public as a fully mobilized legalized program, again with many of these technologies approved for testing of riot/crowd control. Two of the primary contractor companies engaged are DOD contractors, focusing on U.S. aerospace systems are defense suppliers, Lockheed Martin, Northrop Grumman and Raytheon and the computer systems supplier Hewlett Packard supplying the technology for use by military and law enforcement.

Another targeted individual marked for death before possible usage in global perception management operation effort many believe was also Aaron Alexis.

Extremely low frequency (ELF) weapons are part of the weapons systems of most of the modern vessels fielded by the Navy, and other military branches, and Aaron Alexis worked closely in these Navy Operations.

Aaron Alexis Reported:

"I fear the constant bombardment from the extremely low frequency weapon (ELF) is starting to take its toll on my body,"

Aaron Alexis, an extremely low frequency targeted Mind Control Victim reported this just before his rampage.

In reality, MK ULTRA Manchurian, Aaron Alexis was just another individual primed for usage as a patsy who was also a Hewlett Packard contractor, whose life was programmed for destruction by programmed psychopathic minions working these operations who pick and choose whether a life is valueless or valuable to be used as a Manchurian.

Among the many telemetry instruments being used today, are miniature radio transmitters that can be swallowed, carried externally, or surgically implanted in man or animal. They permit the simultaneous study of behavior and physiological functioning. Dr. Carl Sanders, one of the developers of the Intelligence Manned Interface (IMI) biochip, maintains, 'America uses these with military personnel in the Iraq War where they were actually tracked using this particular type of device.' "

Hypnotic intracerebral control of human behavior by implanted ELF microchips is child play compared to 21st Century advancements in the field, and Hewlett Packard is at the lead of the pack.

Memristor research and development is led by Dr. R. (Richard) Stanley Williams. He is research scientist in the field of nanotechnology and a Senior Fellow and the founding director of the Quantum Science Research laboratory at HP.

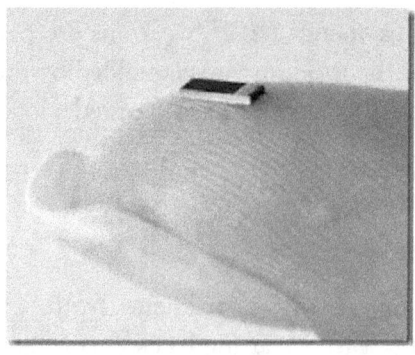

Nanotech Biochip Computer

Nanotechnology (sometimes shortened to "nanotech") is the manipulation of matter on an atomic, molecular and supramolecular scale.

In the nervous system, a synapse is a structure that permits a neuron (or nerve cell) to pass an electrical or chemical signal to another cell (neutral or otherwise). The Memristor is a synapse on a chip making neural computing a reality. In other words, with post-human engineering exploiting discoveries such as the Memristor they will create machines/robots that can learn and think and human brains that could incorporate machines.

Synapse is also a Defense Advanced Research Projects Agency (DARPA) program that aims to develop a nanotech sized electronic neuromorphic machine that scales to biological levels. More simply stated it is an attempt to build a new kind of cognitive computer with similar form, function, and architecture to the mammalian brain. Such artificial brain would be used in robots whose intelligence scales with the size of the neural system in terms of total number of neurons and synapses and their connectivity.

Hewlett Packard Memristor can be described as an agent of transformation conferring robots with the ability to learn (a human trait) thereby rendering them as undecidable, i.e., neither machine nor life. Mirroring its transformative agency in robots, the memristor could also confer the human brain with machine/robot status and undecidability when used for repair or enhancement. The work is part of the DARPA's Synapse Program, or Systems of Neuromorphic Adaptive Plastic Scalable Electronics. Since 2008, the Synapse team (HP) has been developing a new paradigm for "neuromorphic computing" modeled after biology.

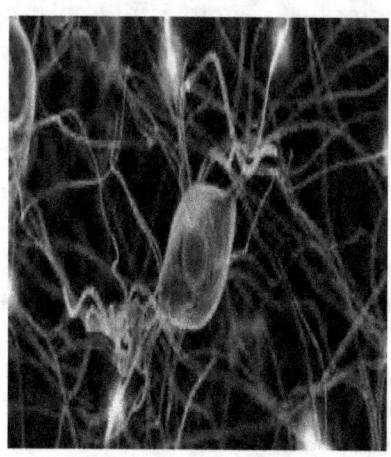

Said to be an example of nanotechnology in the body

In other words, HP has the secret technology and ability to not only track its classified intelligence community private contractors, but control their behavior and physiological functioning through Memristor Nanotechnology like a machine on an atomic, molecular and supramolecular scale. Additionally, it is alleged that every U.S. Senator and Congressional representative has a secret ELF wave length "wavie" file. It seems a logical extension for the Shadow Government to maintain "wavie" files on its Top-Secret Clearance Private Contractors like Michael Dunn and Aaron Alexis to mind control or dispose of them as needed with biometric computer chips, ELF microchips and Memristor Nanotechnologies secretly coupled to central computers. A Memristor chip on the scale of a molecular cell may be impossible to detect without extremely sophisticated detection equipment and computers.

"I have what I believe to be the locations for where they've been developing these weapons for decades …" Aaron Alexis to FFCHS

In the weeks leading up the Naval Yard Massacre, Aaron Alexis was desperately in search of the government central ELF or Nanotechnology computer center to stop his pain. In the process of his frantic search and communications with FFCHS regarding classified government projects, they flipped the switch on Alexis. He became an instant false flag poster boy and patsy for U.S. legislative gun control.

DID THAT SHOTGUN FIT IN THAT BAG?
"Comey said Alexis arrived for work at about 8 a.m. Monday and parked on a deck across a narrow road from Building 197. Carrying a bag containing a Remington 870 shotgun, he entered the building and went to a bathroom on the fourth floor." WASHINGTON POST

"We are told that Aaron cut off the barrel and the stock of the Remington 870 so he could put it in the bag and carry it into the building undetected (no metal detector and x-ray machine in that building?) Yet, if you watch the video of Aaron entering the building, you notice there is no way that shotgun is inside that bag. The bag on his shoulder is shorter than his forearm and that gun certainly isn't. Aside from the fact that the video seems rather staged (obviously it doesn't show the shooting), the gun simply doesn't fit in the bag. "

Alexis' post shooting backpack appears as if it is still full- Alexis appears puzzled about something with the Shotgun

Contrary to news reports, Alexis never had, purchased or used an AR-15 semi-automatic rifle in the Naval Yard Massacre. He didn't have a 9mm Glock that had been conveniently dropped near his body. Inside the building, silent surveillance videos show him ducking and dodging like an actor in a stage play armed only with a Remington 870 shotgun that he didn't seem familiar with and only one shot in the chamber; or he was puzzled and shocked to find that there was no shot at all in the chamber. No witnesses saw Aaron Alexis shoot anyone. A Remington 870 is indeed an odd and strange choice for an armed assault on a highly armed and guarded military government facility. His certainty wasn't trained in speed loading the weapon to kill one person let alone 12 to 13 people.

The Remington 870 is a pump action shotgun. This means that it can hold one shot at a time that you load inside the gun. So, again, where are ammo/shots to mortally wound 12 to 13 people? Aaron Alexis didn't do it!

In regards to Myron May's belief of nanotechnology and my belief of other methods at play, it appears that EVERYTHING is being used today.

See also extensive Mind Control Patents at the tab at:

http://bigbrotherwatching us.com

I don't know how to convince people that PATENTED TECHNOLOGY IS AT PLAY proven by trademark official patents. There are documented patents to create a microwave "Hearing Voices" effect, Neurophone technique, thought deciphering patents, consciousness altering patented technology, and literally hundreds of others that are factually designed to manipulated and influence the target's brain, thought process and perceptions and even dreams.

Vital too many of the patents are the ability to decipher thought and they work in sync with patented EEG cloning software, such as HP precrime software, DHS Malintent. EEG cloning patents can manipulate a target's emotions, by first downloading any emotion, and then beaming fear or panic downloaded in the target's biometric blueprint back to the target at an opportue time. An overwhelming powerful emotion can be used to push a target into tragic action especially if unaware that they are being manipulated and influence or if consciousness is altered.

The Neurophone patent is also documented to also mimic the target's voice so that the target thinks that what is being subliminally projected is heard as their own personal thoughts and the plan or scheme are factually their very own idea. And operation center personnel can also mimic other races voices with this technology.

I would venture to say that of the thousands reported today as "Hearing Voices", that those that are experiencing the technological aspects by technology are well past the noted age of 16 to 26 which psychiatrist agree schizophrenic symptoms typically begin. The problem is that many cannot believe that technology testing is happening on a massive level today. As with many, Myron May showed no symptoms of any type of mental illness, delusions, or psychosis until the age of 31.

I have learned that many targets, especially professionals, will not 100% admit to the technological aspect of the patented hearing voices effect. Admitting this is instant destruction of the target's life and self-undoing. This is especially related to vital employment, and relationships due to strategic misdiagnosis labelling across the board for anyone reporting this technological dynamic. We are not talking one or two people but thousands being deemed schizophrenic strategically when reporting they are in a human guinea technology testing program. This is even after mounds of open literature evidence that factually documents it to be a technological capability and defined as subliminal, synthetic or artificial telepathy, or again, as fondly called by the Department of Defense the "Voice of God".

In my case, I could care less. I say, "Just don't call me late for lunch!" And, by the way, "I am not lunch!"

These technologies are very real and in full use, make no mistake about it. Naturally efforts are intense also to keep this well-hidden evil program from the public to the extent of silencing anyone who could appear credible and a threat. Also, if known the technology can become ineffective in some cases.

I am sure that Nazi trained psychiatrist will eventually began a mass perception management campaign resulting in a new disorder stepped up to include highly educated people after the onset age, so as to continue the cover-up testing, and allowing the program to continue to flourish for behavior modification studies.

The fact is schizophrenics just don't have the presence, structure or organize mental thought process, much less the capability to become exceptional high achievers as attorneys, ex-FBI agents as well as many from all walks of life who have led normal lives until targeted.

The books in this series were well researched then the information applied, formatted, and produced in a comprehensive manner reveal this fact alone. In my case, as many targets agree, we will push these unpatriotic, pathetic, psychopathic souls working these secret programs to suicide before they push us!

There have been several efforts to leave the impression of me as mentally ill and strategically promoted related to me. In Google Images surrounding my book promotions, I document also, as shown below, that the below image miraculously appeared out of the blue. I document also, in another book, the fact that when I attempted to delete, it several times, it was continually reposted. I then decided to go with it and added one of my book images and my comments at the top and bottom of the posting as shown below.

NOTE: The link to the Jesse Ventura, Brain Invaders episode can be viewed at link:

http://tv.naturalnews.com/v.asp?v=73596F521ABC983504AF22E59019FF18

Many know of also the reality of covert stalking as factual. This is especially true for those recruited in communities, nationwide, then silenced under national security secrecy orders related to the technology in use specifically. This could be interpreted as intentional efforts unethically deny even basic human rights to US citizens.

MK ULTRA Manchurian Candidate Aaron Alexis, 34 years old, as well as Myron May, by all reports, were both "nice guys". Again, Alexis was a U.S. Department of Defense Computer Consultant, and a Florida based computer systems specialist subcontracted by a company called "The Experts," which is a subsidiary of HP Enterprise Services and is owned by Hewlett Packard.

"On September 16, 2013 in what many believe to be another false flag operation, allegedly Alexis mass murdered 12-13 people in Washington DC. Alexis had been a confident and skilled computer and network installation specialist. He did the extensive hands-on ground work connecting the complex vast system network of HP wires and hardware which tie the NSA and military intelligence listening and data collection centers.

It has been reported that Alexis had a role in transferring HP's global network systems vital to NSA intercepts of satellite communications from NSA facilities at East Coast naval bases to Colorado.

On 9-11, Explosions were reportedly going off in the Twin Towers underground basements before the planes hit the Towers.

On 9-11, Alexis is also believed to have been in the one of the Twin Towers' underground/basement telecommunication wiring center prior to the collapses. According to witnesses, explosives were setoff in the twin tower basements just before the guided missile planes hit the buildings. An argument has been raised that he was one of the necessary hands on technicians in the 9-11 Deception/Inside Job. He would have been one of the essential technicians that managed the wiring and hardwire necessary to synchronize explosives with guided missile planes, and the timing of the controlled demolition of the Twin Towers and Building 7.

Secretly, Alexis had worn three hats for the shadow government. He worked for the National Security Agency (NSA), CIA and U.S. Navy Reserves – thereby earning a sizeable pay package that allowed extended visits to Thailand and plush accommodations wherever he went until the music stopped.

Weeks before the shootings, both Alexis and May contacted the targeted individual support group, Freedom from Covert Harassment and Stalking (FFCHS) for help. Alexis told the group that the Navy was electronically assaulting him with an ELF weapon, and he was desperate to protect himself. May told several people again of being relentlessly hit by "Direct" Energy Weapons which could have been drone deployed and biometrically authenticated

(Above excerpts from mindcontrolblackassassins.com.)

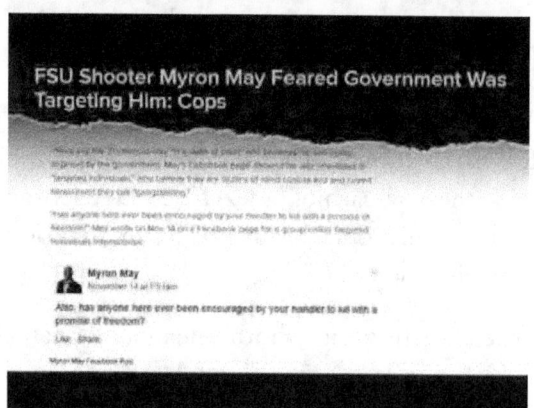

November 14, 2014 Facebook post appearing to be a strategic omen for Myron May.

This is what Myron May factually wrote on the Targeted Individuals International Facebook page, on November 14, 2014, when he sought to connect himself with the community of activist.

The Targeting of Myron May

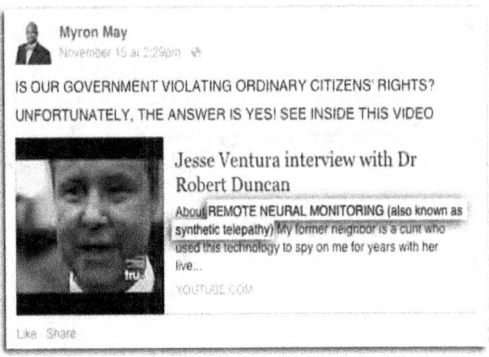

If you Google Targeted Individuals you will be met with credible experiences of thousands of highly educated individuals, who continue to substantiate what is happening today and who continue to endure this program. This is although many have reported subliminal suggestions to commit suicide as well.

CHAPTER SEVEN
A Life Discredited and Destroyed

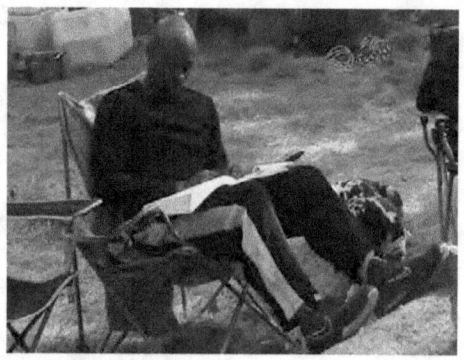

Myron May at a camp out with family friends the Tauntons. May can be seen devoting time to studying for the Florida BAR exam during the outing.

How could, by all accounts, of those who knew Myron May personally, or who worked with him, go from this to what resulted in yet another loss of life and injury by someone everyone believed had a very promising future?

Media news report of Shooting at Strozier Library by Tallahassee Police Department saying:

"Mr. May had a written journal and videos where he expressed fears on being "targeted," and he wanted to bring attention to this issue."

The Targeting of Myron May

Myron May's fraternity nickname was 'sensitive Joe' and it was fitting" state Rep. Matt Gaetz, who belonged to the same Florida State political club as May stated. "I was so surprised that someone with this docile nature would have something happen in their lives that would have this outcome.

Many within the Department of Defense and the Association of psychiatry know that hearing voices by many technological methods, to include appearing to be coming from the walls is not an unrealistic technological capability today.

"Remote Neural Monitoring" is a term coined by one of the first targeted individuals hoping to bring a lawsuit, in 1992, against the National Security Agency, contractor, James Akwei, also African American.

If anyone hasn't heard the story, John Akwei writes: I authored precise details of NSA Remote Neural Monitoring during the summer of 1992, after Civil Action 92-0449 was dismissed at the DC Federal District Courthouse.

He visited 22 libraries in order to research this information, relying a lot on the InfoTrack database, (before the Web). I then distributed this document to a lot of public law non-profits, and one of them published this document to the Web in 1994.

During the next 20 years a lot of alterations appeared, therefore I am grateful that what I originally wrote had been restored says John Akwei.

Many in the Psychiatric field shamelessly are well aware of the microwave technological "Hearing Voices" effect and that this technological capability factually exists which can isolate and bean voices directly into a single person's head.

History documents that the U.S. Association of Psychiatry were heavily involved in testing of this technology dating back decades for Remote Neural studies and in combined efforts the Canadian psychiatric association. These were ongoing Nazi style mind control efforts of which I document in the history of these programs in "Remote Brain Targeting."

Case and Point: In 1975, a neuropsychologist Don R. Justesen, the director of Laboratories of Experimental Neuropsychology at Veterans Administration Hospital in Kansas City, unwittingly leaked National Security Information. He published an article in "American Psychologist" on the influence of microwaves on living creatures' behavior.

In the article he quoted the results of an experiment described to him by his colleague, Joseph C. Sharp, who was working on Pandora, a secret project of the American Navy.

Don R. Justesen wrote in his article:

"By radiating themselves with these 'voice modulated' microwaves, Sharp and Grove were readily able to hear, identify, and distinguish among the 9 words. The sounds heard were not unlike those emitted by persons with artificial larynxes" (pg. 396).

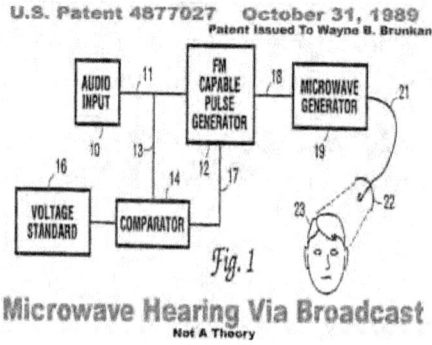

Microwave Hearing Via Broadcast
Not A Theory

In fact, here is a factual patent in detail describing a version of this technology which is also patented at the United States Patent and Trademark Office. It is accurately described as a thought transmission unit which sends modulated electromagnetic wave beams to a human receiver to influence thoughts and actions without an electronic receiver.

US Patent DE 10253433 A1

Abstract:

A thought transmission unit sends modulated electromagnetic wave beams over long distances to a human receiver to influence the thoughts, actions or perceptions of the organism with or without their consent but without them requiring an electronic receiver.

Claims (33) translated from German

1. Radio means, characterized in that - the radio device generates bundled modulated electromagnetic radiation and transmits to a human recipient, - the carrier frequency of the combined modulated electromagnetic radiation of between 10 6 Hz (= 1 MHz) and 10 14 Hz (= 100 THz) is located, - a modulation frequency of the carrier frequency is between 0,01 Hz and 10 11 Hz (= 100 GHz) is located, - the distance between the radio relay apparatus and the receiver is more than 10 m, - acting the combined modulated electromagnetic radiation onto the body of the receiver in such a way that a substantial likelihood that a proposed change in the thoughts or actions of the receiver is created - changing the thoughts or actions of the receiver is detectable using scientific methods, - the mission of the bundled modulated electromagnetic radiation from the receiver itself is not consciously perceived, - the information content of the consignment of the bundled modulated electromagnetic radiation comprises more than 100 bits, - the receiver for receiving data transmitted by the combined modulated electromagnetic radiation information no electronic aids required, the conversion of electromagnetic radiation into acoustic or optical or mechanical signals, or odor signals or taste signals cause.

2. Radio relay device according to claim 1, characterized in that additionally also aware perceptible signals sent by the receiver.

3. Radio relay device according to claim 1 or 2, characterized in that the transmission of the bundled-modulated electromagnetic radiation at the receiver at least one of the following five based on the action of the electromagnetic radiation effects caused by: (i) underlying signals in the range of 12 Hz - 25 kHz, (ii) perceptible signals in the range of 12 Hz - 25 kHz, (iii) underlying signals with frequencies below 12 Hz, (iv) underlying signals with frequencies above 25 kHz, (v) detectable signals at frequencies outside the range of 12 Hz - 25 kHz.

4. Radio relay device according to claim 1, characterized in that sounds of a language is converted into a sequence of pulses and said sequence of electromagnetic radiation is modulated.

5. Radio relay device according to claim 1, characterized in that it involves a camera or other detection means sensitive in the carrier frequency of the modulated electromagnetic radiation for the bundled transmission of thought.

6. Radio relay device according to claim 1, characterized in that it involves a computer, which calculates to send stimuli for an intended telepathy.

7. Radio relay device according to claim 1, characterized in that the combined modulated electromagnetic radiation by more than 50% derived from a source of emission of radiation induced.

8. Radio relay device according to claim 1, characterized in that the combined modulated electromagnetic radiation by more than 50% from a maser laser phased array, diode bundle comes magnetron or klystron.

9. Radio relay device according to claim 1, characterized in that the combined modulated electromagnetic radiation by more than 50% in less than 1 degree x 1 degree solid angle radiated.

10. Radio relay device according to claim 1, characterized in that the distance between the radio relay apparatus and the receiver is greater than 1 km.

11. Radio relay device according to claim 1, characterized in that the combined modulated electromagnetic radiation applied to the recipient organism in such a way that with more than 5% of an intended change in the probability thought or action of the receiver is generated.

12. Radio relay device according to claim 1, characterized in that the combined modulated electromagnetic radiation applied to the recipient organism in such a way that with more than 95% probability that an intentional change of thought or action of the receiver is generated.

13. Microwave process, characterized in that: - the bundled modulated electromagnetic radiation generated and transmitted to a human recipient, - the carrier frequency of the combined modulated electromagnetic radiation of between 10 6 Hz (= 1 MHz) and 10 14 Hz (= 100 THz) is located, - a Modulation frequency of the carrier frequency is between 0,01 Hz and 10 11 Hz (= 100 GHz) is located, - the distance between the radio relay apparatus and the receiver is more than 10 m, - the combined modulated electromagnetic radiation applied to the body of the receiver in such a way that a substantial likelihood that a proposed change in the thoughts or actions of the receiver is created - changing the thoughts or actions of the receiver with scientific methods can be proven - the mission of the bundled modulated electromagnetic radiation from the receiver itself is not consciously perceived, - the information content the mission of the bundled modulated electromagnetic radiation comprises more than 100 bits, - the receiver for receiving data transmitted by the combined modulated electromagnetic radiation information no electronic aids needed, causing a conversion of electromagnetic radiation

into acoustic or optical or mechanical signals, or odor signals or taste signals .

14. Radio relay method according to claim 13, characterized in that additionally to the recipient is aware of perceptible signals.

15. Radio relay method according to claim 13 or 14, characterized in that the transmission of the bundled-modulated electromagnetic radiation at the receiver at least causes one of the following five based on the action of the electromagnetic radiation effects: (i) underlying signals in the range of 12 Hz - 25 kHz, (ii) perceptible signals in the range of 12 Hz - 25 kHz, (iii) underlying signals with frequencies below 12 Hz, (iv) underlying signals with frequencies above 25 kHz, (v) detectable signals at frequencies outside the range of 12 Hz - 25 kHz.

16. Radio relay method according to claim 13, characterized in that sounds of a language is converted into a sequence of pulses and said sequence of electromagnetic radiation is modulated.

17. Radio relay method according to claim 13, characterized in that it involves a camera or other detection means sensitive in the carrier frequency of the modulated electromagnetic radiation for the bundled transmission of thought.

18. Radio relay method according to claim 13, characterized in that it involves a computer, which calculates to be transmitted to an intended transmission of thought stimuli.

19. Radio relay method according to claim 13, characterized in that the combined modulated electromagnetic radiation for more than 50%, derived from a source of emission of radiation induced.

20. Radio relay method according to claim 13, characterized in that the combined modulated electromagnetic radiation by more than 50% from a maser laser phased array, diode bundle comes magnetron or klystron.

21. Radio relay method according to claim 13, characterized in that the combined modulated electromagnetic radiation by more than 50% in less than 1degree x 1 degree solid angle radiated.

22. Radio relay method according to claim 13, characterized in that the distance between the radio relay apparatus and the receiver is greater than 1 km.

23. Radio relay method according to claim 13, characterized in that the combined modulated electromagnetic radiation applied to the recipient organism in such a way that with more than 5% of an intended change in the probability thought or action of the receiver is generated.

24. Radio relay method according to claim 13, characterized in that the combined modulated electromagnetic radiation applied to the recipient organism in such a way that with more than 95% probability that an intentional change of thought or action of the receiver is generated.

25. Radio relay method according to claim 13, characterized in that based on the effect of modulated microwave energy sense interference is involved.

26. Radio relay method according to claim 13, characterized in that transmission is thought to a target person from items of concrete, stone, plastic, or wood therethrough.

27. Microwave method according to claim 13, characterized in that telepathy is a target entity has more than 10 kilometers away.

28. Radio relay method according to claim 13, characterized in that the receiver uses a means for amplifying the signal thought, for example, a microwave antenna or a booster.

29. Radio relay method according to claim 13, characterized in that the receiver from the transmitter is observed by a camera, and the carrier frequency of the modulated electromagnetic radiation for the bundled transmission of thought is a

frequency at which the camera for observing the receiver is sensitive.

30. Radio relay method according to claim 13, characterized in that the transmission of signals necessary for the intended idea-based computer predicted by utilizing correlation between a set of stimuli and responses.

31. Microwave method according to claim 13, characterized in that the transmission of thought goes directly from sender to receiver or a Bündlungseinrichtung or amplifier device or relay station.

32. Radio relay method according to claim 13, characterized in that the carrier frequency is modulated to an intermediate frequency, which is modulated to the desired signal.

33. Radio relay method according to claim 13, characterized gekeennzeichnet that words in pulse trains whose envelope corresponds to the intensity variation of the words formed and computer-stored and the computer accessed pulse trains modulated on the electromagnetic beam and are transmitted at such low intensities, that the receiver the transmission is not perceived consciously.

Description translated from German

- Background of the Invention
- Field of the Invention
- [0001]

The invention relates to long-range and long-range telepathy mind reading. Applications include the extension of the traditional means of communication, support important public appearances and personalities important negotiations, the mission of major hazard information in emergency situations, the active avoidance of significant risks, the study of criminals who support brain research. Here are limitations of conventional methods of information transfer, such as mobile phone, radio and television, overcome.

- Description of the prior art

- [0002]

In modern media, such as radio and television, an electronic device is needed that converts electromagnetic radiation into a detectable acoustic or optical signal and to provide individuals are not generally individually with information. The effect of the audibility of certain RADAR pulses (observations during World War II), (b) is also known as (a) the direct acoustic visibility ("audibility") of modulated microwave energy with irradiation in the head (1 , Frey, 1961, Frey, 1962, Frey & premeasured, 1973, Lin, 1978; Frey & Corin, 1979; Brunkan, 1989, Lin, 1989; Stocklin, 1989; Frey, 1993), (c) the feeling of control by acoustic or electrical stimulation (Meland, 1980; Gall, 1994), and (d) the use of acoustic signals to subliminal (Lowery, 1992). The acoustic perception by the action of pulsed microwave radiation have been selected based on the majority of experimental conditions on the generation of thermoelastic pressure waves in the inner ear (Lin, 1989).

- [0003]

The human Körperdipol has a resonance frequency of 80 MHz at 1.80 m length. The individual something different electromagnetic resonance frequencies of the human head will be around 400 MHz to 700 MHz in adults and in infants (Lin, 1989). Because of the skin effect penetration depth of the electromagnetic radiation into the organism is dependent on frequency, such as at an irradiance on the head is the absorption at 2.5 GHz frequency mainly in the outer 1-2 cm of the brain against more at 900 MHz within the brain (Lin, 1989).

- [0004]

Are also known electromagnetic weapons which can be stunned or off (when observed by millimeter-wave telescopes or microwave detectors) over long distances or through non-metallic walls through people.

- [0005]

It is also known that subliminal stimulation with conventional acoustic methods. For example, modulated rhythms at 1.7 - 3.5 Hz are used to promote sleepiness. Abnormal states of consciousness can be described by rhythms in the 3.5 to 7 Hz and 28 - 56 Hz promote. The normal rhythm of the human brain is 7-14 Hz and 14-28 Hz in the case of arousal or anxiety (Gall, 1994).

- [0006]

The ideas of telepathy and mind reading are however usually not considered viable fantasies (see, e.g., Chapman, 1998) and none of the mentioned systems alone can be efficiently realized long-range telepathy or even mind-reading, for example, over a distance of several kilometers. People who claim that without technical aids send thoughts over long distances or to receive (e.g. some esoteric), have been able to lead any evidence of effect. Also supported by numerous utopian films with far-reaching episodes of telepathy or mind reading that there has been this dream is not a viable solution with good efficiency.

- Objective of the invention

- [0007]

The aim of the invention is to expand the possibilities of modern media in the form of long-range transmission of thought on the part of the recipient no electronic aids such as radio, television or mobile phone needs.

- Literature

- Brunkan, WB (1989) Hearing System. U.S. Patent 4,877,027.

- Chapman, RK (1998) Mental telepathy debunked: counter-arguments against the concept of thought transmission, and mind-reading ideas. = ISBN 0-9698637-6-4.

- Frey, AH (1961) Auditory system response to modulated electromagnetic energy. Aerospace Med 32, 1140-1142.

- Frey, AH (1962) Human auditory system response to modulated electromagnetic energy. J. Appl. Physiol. 17, 689-692.

- Frey, AH & reasonable, R. (1973) Human perception of illumination with pulsed UHF electromagnetic energy, Science 181, 356-358.

- Frey, AH & Corin, E. (1979) Holographic assessment of a hypothesized microwave hearing mechanism. Science 206, 232-234.

- Frey, AH (1993) Electromagnetic field interactions with biological systems. FASEB Journal 7, 272-281.

- Gall, J. (1994) Method and system for old ring consciousness. U.S. Patent 5,289,438, and references therein.

- Lin, JC (1978) Microwave auditory effects and applications. Charles C. Thomas, Publisher, Springfield, IL.

- Lin, JC (1989) Electromagnetic interaction with biological systems. Plenum Press, New York.

- Lowery, OM (1992) Silent subliminal presentation System. U.S. Patent 5,159,703.

- o Meland, BC (1980) Apparatus for electro physiological stimulation. U.S. Patent 4,227,516.,

- o Stocklin, PL (1989) Hearing device. U.S. Patent 4,858,612.

- Detailed Description of the Invention

- [0008]

The invention is based, to provide particular desired message transfers without being subject to the limitations of commonly used electronic means the task. According to the invention the object is achieved by the use of far-reaching thought transmission, telepathy based on radio. Unlike conventional radio, however, the electromagnetic beam (thoughts) beam is injected directly into the body of the recipient, such as the head, the cerebral cortex, the inner ear, the auditory nerve or optic nerve. Depending on particular introduced into the electromagnetic beam signals (for example, by amplitude modulation), this coupling causes the receiver an intention to change of mind. In general, the change of mind of the receiver is only statistically effective, i.e. it will only increase or decrease the probability of certain thoughts on unintended ways. However, in individual cases, the change can be determined. The telepathy is suitable in some applications to be combined with observations using millimeter wave cameras and microwave-based voice transmissions, in which the audibility is used by modulated microwave energy, but can also be operated independently.

- [0009]

For example, in a simple embodiment of a thought transmission device says the operator of the device (observer Observer) to be transmitted thoughts in a microphone, the microphone, the electric signal is converted into a sequence of pulses by means of an electronic (for example, square pulses of

100 microseconds duration 200 microseconds distance, where appropriate pulse sequences are computer-stored and retrieved from there as needed), the sequence of pulses is your microwave beam is modulated, which is sent to the recipient and such a low intensity is that the receiver has no conscious awareness of the program, but this only subliminally acts. Instead of the pulse train signal a flankenversteilertes (e.g. by means of multiple squaring) or the original signal can be used.

- [0010]

  For example, in a more complicated version of telepathy device gives the operator (Observer, Observer) of the device to be sending thoughts into a computer (or other carriers), which translated using tables or neural networks to be sending thoughts into a sequence of signals, which is the microwave beam which is transmitted to the receiver is modulated. This sequence of signals may contain microwave-induced consciously perceivable acoustic signals (e.g., clicking sounds, rhythms, voice, music) and microwave induced only unconsciously perceived acoustic signals (e.g., clicking sounds, rhythms, voice, music) and microwave induced low effective electrical rhythms. The calculation of the translation tables to be sent between thought and sequence of signals is carried out e.g. by utilizing a set of correlations between stimuli and responses. The training of the neural networks to be transmitted to the gear ratio between thought and sequence of such signals is done by observing the response to a set of stimuli.

- Frequencies

- [0011]

  For better effectiveness of telepathy certain resonances of body parts (such as the head, parts of the inner ear, optic nerve) can be chosen as the carrier frequency of the electromagnetic beam or modulated onto the carrier frequency frequencies. For example, carrier frequencies, and may not be suitable carrier frequency modulated at frequencies between 80 MHz and 400-700 MHz and 1 - 100 GHz for addressing the body or the head or parts of organs (e.g. inner ear, nerves).

- [0012]

1 GHz - for modulated signals mainly the frequency range of 1 Hz is. For example, the frequencies for speech signals (consciously or subliminally perceptible effective) in the region of 16 Hz - 20 KHz, but as far when transformed into a pulse train about, for example in the MHz range. Extremely low frequencies are suitable for example to influence the state of consciousness and feeling influence. For example, in analogy to the traditional acoustic stimulation modulated rhythms at 1.7 to 3.5 Hz and from 3.5 to 7 Hz and 28 - 56 Hz are used to promote sleepiness or altered states of consciousness.

- Modulation
- [0013]

For the modulation of the electromagnetic beam, there are various options that can be used individually or in combination, e.g. (a) sounds of a language or other signals are converted into a pulse sequence that is modulated onto your electromagnetic beam, or (b) sounds of a language or other signals are directly modulated on the electromagnetic beam. The show is perceptible or imperceptible - depending on e.g. intensity, type of modulation, place of exposure to the organism and frequency.

- Radiation sources
- [0014]

For the generation of the electromagnetic beam (thoughts) beam are especially maser (Microwave Amplification by Stimulated Emission of Radiation) and LASER (Light

Amplification by Stimulated Emission of Radiation), where the wavelength is not necessarily in the classical microwave range (300 MHz - 300 GHz) must be (2). According to the invention, radiation sources which involve stimulated emission of electromagnetic radiation that lies outside of the conventional microwave region involved. In particular maser are at all points of the "Detailed Description of the Invention", the embodiments and figures, including legends synonymous with induced emission of radiation sources, such as maser and laser (eg, Free-Electron Laser). Further suitable radiation sources, magnetrons, gyrotrons, klystrons, semiconductor diodes and phased arrays into consideration.

- Transmission power
- [0015]

Depending on the size of the transmission losses, the transmission power per addressed person in the power of the human brain can (approx. 40 W) located or slightly higher, but also to be significantly lower in e.g. exposure to the inner ear or even the nerve endings of the senses, depending on the application. For example, in transmissions over several kilometers through building walls and through transmission power of about 1000 W per person addressed may be needed to compensate for the transmission losses. Special measures may be necessary to prevent weapons effect on individuals in the beam path (a high-energy radiation can stun and short-term temperature increase of the brain above 45 ° C can be fatal). On the other hand, in the absence of significant transmission losses, a power of 1 W is much less than average for a subliminal telepathy enough. Since excessive absorption of microwaves in tissue can cause damage (especially in rapidly dividing cells and neurons), you will prefer low total radiation energy in many applications.

- Automation

- [0016]

The thoughts can be transferred from the example telepathy device-to-human or semi-automatically or fully automatically from person to person with intermediate telepathy device.

- 1 Embodiment

- [0017]

Mounted on a vehicle telepathy device that produces a focused microwave beam is modulated in an appropriate manner and to the receiver (target) sends (3). The total weight of telepathy device with maser (maser) for generating the electromagnetic beam (Beam), microphone can be used for inputting voice signals by the observer (headset), rechargeable energy source to buffering of current fluctuations and detector (detector) to monitor and support the beam tracking e.g. 100 are kg. In order to allow for the electromagnetic beam at a good Bündelbarkeit still sufficient penetration of the air, walls, and soil can be used as for example, the carrier frequency of the maser range 1-1000 GHz. The carrier frequency of the electromagnetic beam (beam) for transmission of thought can be a frequency at which the detector (detector) for monitoring the receiver is sensitive, for example. Transmission of thought and observation of the recipient e.g. via long distances by air or by walls of concrete, stone, plastic or wood.

- [0018]

The idea is transmitted as by the observer directs the beam maser (beam) on the head of the recipient (target) and speaks into the microphone, with the microphone, the electrical signal by the electronics of telepathy device, the carrier frequency of the maser in an appropriate manner (e.g., in the form of a pulse train whose amplitude is correlated to the amplitude of the microphone the electrical signal) is modulated and wherein the maser radiation induced voltages in the head of the receiver, which acts as a signal at the receiver as subliminal. Alternatively, the modulation of the microphone signal to the carrier frequency can be carried out for example using an electronic transmission device which has been previously trained for example taking advantage of a correlation between the set of stimuli and responses. Alternatively, the sequence thought to be sent are input into a computer, which calculates the signal to be transmitted. To develop the program for the translation to be sent into the thoughts of the electromagnetic beam (Beam) aufzumodulierenden sequences may have been used as a large number of weak correlations between thoughts and stimuli. The computer program may, for example, a neural network (4) Included, which has been previously trained with such a large set of pairs of stimuli and reaction and thoughts projected intention to sets of stimuli after training. The sequences need not be modulated in the audible frequency range. For example, also low-frequency signals in the range 1-20 Hz, may be modulated onto the carrier frequency of the maser, which can result in a loss of the recipient etc. Also, signals in the range above 20 kHz may be used. In many cases - especially if you do not want to engage in the very acts independent of the receiver - will be tuned to be content with an unobtrusive and unconscious thoughts of the receiver changes the probabilities.

- 2 Embodiment

- [0019]

A handheld telepathy device a maser (maser), a microphone (headset) to the input of the speech signals by the observer (Observer), a rechargeable energy source (battery) and monitor a detector (detector), for example, a millimeter wave camera contains (5). The telepathy device can be connected to the electricity fixed, the power supply of a vehicle or a generator (power generator) with e.g. 200 W power. By the display (display) and the handle (handle) is tracked, the idea transmitter to the receiver (target). Various switches (switches) and electronics (Electronics) allow you to set modes such as transmission of stored signals, automatic intensity adjustment, type of modulation to transfer the voice signals of the observer. The telepathy device can be flexibly mounted on tripods or vehicles by a connecting member (Connector). Telepathy for example, by the broadcast of previously determined sequences. Unlike numbing shots with electromagnetic weapons will work with relatively low intensities. Intensity below the conscious perception affects the electromagnetic radiation to which such an acoustic signal is modulated, seemingly unconscious as subliminal acoustic signal and influences the thoughts of the receiver. At higher intensities of the electromagnetic beam is directly felt. In addition to audible and subliminal language, music and rhythms can be as modulate the electromagnetic beam and low frequency rhythms (e.g., 16 Hz) and signals in the range above 20 kHz.

- 3 Embodiment

- [0020]

To a vehicle (6), A transmission tower (7), A house (8) Or a flying object (9) Mounted (possibly motion-stabilized) telepathy device with a source of intense electromagnetic radiation and means for modulation of the radiation corresponding to the to send thoughts, such as a computer for to send thoughts a sequence of electromagnetic stimuli calculated (e.g., subliminally or consciously perceived language music, rhythms and sound sequences that are sent simultaneously or sequentially). For telepathy using near-Earth satellites (10), The maser (maser) to a very small opening angle. Ideas are generated, for example by taking advantage of many weak correlations between thoughts and sentences of stimuli. When used over a long period a large number of correlations can be measured and can exploit the relatively weak correlations between stimuli and thoughts to a significant change in the probability result of certain thoughts. To be used for highly collimated radio carrier frequencies suitable for high transmission of low thought signals, signal to be transmitted is modulated onto the carrier frequency of the radio beam, for example by amplitude modulation. If the modulated signal is an audible signal (such as an amplitude modulation with an audible frequency occurs), above a certain intensity of this modulated electromagnetic radiation can be heard directly as a seemingly acoustic signal. To reduce the need for thoughts transmission intensities of electromagnetic radiation, irradiation can use in individual nerve bundles such as auditory and optic nerves. This can not only occur via the use of the resonance frequencies, but also by irradiation with high precision so that the body parts are preferably made by the beam.

- 4 Embodiment
- [0021]

Thought transmission to recipients in case of disaster (11). Telepathy can be useful in situations important exception Mitigation and uncomplicated control of fast rescue. Subcomponents of telepathy can be microwave-assisted voice transmissions and feel influences the receiver. Stimuli can be as language, music, rhythms and sound effects. The stimulation can subliminally (i.e. unconsciously perceived) or consciously perceived. Several stimuli may be sent simultaneously or sequentially, in order to trigger a particular response. For example, the broadcast consciously perceivable parts of words with the mission of acting subliminal rhythms are combined. Telepathy for example, has proposed a change in the mind of the recipient, such as motivation to mitigating actions to follow.

- 5 Embodiment
- [0022]

Human-to-human transmission of thought: The signal to be transmitted is taken directly from a person's head and directly or in processed form (e.g. by means of frequency analysis and selection of the dominant frequency) modulated onto the electromagnetic beam. In this manner are, for example, stress or relaxation states are distinguished by different frequencies of brain activity transmitted. Transmitter or receiver can be, for example people in a vegetative state or deaf-blind.

- 6 Embodiment
- [0023]

Profiling and mind reading with a convicted criminal under the law and morally permissible. A simple method would be surprising to subliminally send the person a keyword that only has important meaning for them and by simultaneous observation of the reaction, a suspicion is reinforced or softened. The mission of the keyword can be a preparatory phase (sensitization phase) preceded the person's thoughts are directed through subliminal signals to the key event in the example. However, the computerized telepathy allows far more sophisticated methods: for example, certain key information can be sent subliminal over a longer period of changing intensities and the reactions of the receiver are correlated with the signal.

- 7 Embodiment
- [0024]

Unobtrusively manipulate or snoop on convicted criminals to avert dangers - as far as legally and morally acceptable (Fig. 12). The temporary off all criminals through intense amplitude modulated microwave radiation at the storming of an object (unobtrusively through walls) has some risk of failure and is difficult in electromagnetically shielded objects. The telepathy possible to reduce these risks. In life threatening situations it may be acceptable to extend a thought manipulation on non-criminals involved people, which simplifies the use of screened objects (e.g. diffuse radiation through holes in the Abschrimung throughout the interior). For example, the radiation power for telepathy less than 1/1000 of the need for the temporary stunning of criminal radiation power may be, which should also be a significant cost factor. Another advantage is that the hardware of telepathy can be easily extended to the microwave-assisted interception of conversations of criminals.

- 8 Embodiment
- [0025]

Brain research and treatment of diseases. The methods, telepathy presented permit new ways of analysis, therapy and prophylaxis of certain pathological disturbances of brain metabolism and influence of certain non-pathological limitations, stress situations and aging of brain metabolism. For example, since the electromagnetic radiation may affect other organ parts as in the application of sound or visible light, opens up new possibilities. For example, in diseases other types of effects can be made to certain neurological processes compared to acoustic stimuli that are not based on the interaction of electromagnetic radiation. Telepathy may also be supportive in molecular medicine e.g. for the analysis of biochemical networks in the brain. In some such applications it may be advantageous to realize the transmission of thought over a few millimeters away.

- 9 Embodiment
- [0026]

Support of negotiations and talks by important people: for example, the presentation of the important person is tracked by a team that can engage an advisory by telepathy. At crucial points of the presentation can be as important thoughts interspersed. With subliminal telepathy is the speaker - in contrast to conventional acoustic transmission via earphones - not disturbed by the transfer.

- 10th Embodiment

- [0027]

For example, the determination of the specific idea for generating electromagnetic signals to be sent in the examples 1-9: The measurements are carried out a large number of correlations between stimuli and thoughts or induced reactions. These correlations are mathematically combined to produce computer-based sequences of stimuli may correlate better with the desired thoughts or reactions. For example, if 100 independent stimuli cause one of the 2% probability of a particular thought, you can combine an approximately 87% probability of a particular idea cause. Since many of the methods of transmission of thought presented are applicable for months, it is impractical in many cases to use relatively weak correlations to obtain a significant result.

- Figure descriptions

- [0028]

1 Relative intensity (1) As a function of frequency in GHz (2), Which is necessary under certain experimental conditions in order to perceive pulse-modulated microwave energy acoustically. (According to data from Lin, JC (1978) Microwave Auditory Effects and Applications. Charles C. Thomas, Publisher, Springfield, IL, USA). At high frequencies, the skin depth decreases in the head, which can lead to sensitivity reduction (prior art).

- [0029]

2 Example of a section of a high-frequency amplitude-modulated carrier signal. The high-frequency radiation, e.g. in the range 1-1000 GHz, can be brought into sharp focus and propagates almost linearly. The envelope curve of the signal shown corresponds to a low frequency useful signal (eg 0.1 Hz - 1 MHz), which for example in the cerebral cortex, in the inner ear or in other organs is effective.

- [0030]

3 Thought transmission to a receiver (3) Using a modulated beam of millimeter waves or microwaves (4), One of the masers (5), e.g. free-electron maser, emanates, which is based on a reconnaissance vehicle (6) Is mounted, for example by means of a stand (7). The maser can as a free-electron maser its (often referred to as Free-Electron Laser). In one mode, speech signals of an observer, for example by means of a microphone (8) Entered in the reconnaissance vehicle maser beam directly amplitude-modulated. In addition, bewußtseinsmodifizierende signals can be modulated. Bewußtseinsmodifizierende Such signals include audible noise that can trigger certain reactions conscious or subliminal sounds audible frequency that can trigger certain reactions unconsciously, or low non-audible signals (modulated electromagnetic beam on the infrasound). The observer can, for example by means of the detector (9), Such as a telescope or a millimeter-wave radar detector, the beam tracking and response of the receiver (3) Track. The combination of the detector (9) And computers (computers with ADC card, amplifier and battery (10) Display (11) Keyboard (12); Joystick (13) Floppy Disc Drive (14), Switches (15)) Adjusts automatically depending on the distance differences and absorbing walls, trees or mounds of the intensity. The computer is using such a cable (cable for connection to the power supply (16)) To a power supply and, for example by means of a cable bundle (17) To a stepper motor mechanics (joint and step motors (18)) Connected to the

beam tracking. Depending on the choice of intensity of the electromagnetic beam (4), Whose modulation and type of action is the transmission of thought to the recipient (3) Consciously or unconsciously. The range of telepathy is, e.g. 5 m - 20 km (19).

- [0031]

4 For example, a neural network (20) To calculate the sets of signals (stimuli (21)) To the production of certain thoughts (reactions (22)) Are sent. The sets of signals are neural nodes that correspond to specific transfer functions, with the idea to send (reactions (22)) Connected.

- [0032]

5 A handheld telepathy device a maser (5), A microphone (8) For inputting the voice signals by the observer (23), A rechargeable power source (accumulator (24)) And for monitoring a detector (9), For example a millimeter wave camera contains. The telepathy device can be connected to the current fixed network, the power grid of a vehicle or a generator (alternator (25)) Are connected with e.g. 200 W power. By the display (11) And the handle (26), The electromagnetic beam (4) Of telepathy device to the receiver (3) Tracked. Various switches (15) And the electronics (27) Allow you to set modes such as transmission of stored signals, automatic intensity adjustment, type of modulation to transfer the voice signals of the observer. The transmission device can transmit thought by a connecting element (28) Are movably mounted on tripods or vehicles. The range of the transmission of thought and observation of e.g. 5 m - 5000 m (29).

- [0033]

6 Thought transmission to a receiver (3) Using a modulated beam of millimeter waves or microwaves (4) Extending from a phased array (30) Assumes that a reconnaissance vehicle (6) Is

mounted. For example, computer-stored words are converted into pulse trains by the computer whose envelope corresponds to the intensity variation of the words, and then the pulse trains modulated on the electromagnetic beam transmitted at such low intensities, and that the receiver (3) The transfer does not consciously perceive. The beam tracking done e.g. by use of the recipient (3) Radiation reflected by the radar principle. The range of telepathy is, for example 10 m - 1000 m (31).

- [0034]

    7 Thought transmission to a receiver (3) By means of the beam (4) A phased array (30) Through a reinforced concrete wall (32) With simultaneous monitoring of the receiver (3) Using millimeter wave camera (33). Steel mesh and small metal objects in the path set due to the conical geometry of the beam is not a significant problem. telepathy device and millimeter wave camera as a tower (34) Is mounted. The range of the transmission of thought and observation of e.g. 50 m - 5 km (35).

- [0035]

    8 Thought transmission to a receiver (3) By means of the beam (4) Of a building (36) Mounted maser (5) With simultaneous monitoring of the receiver (3) By means of a detector (camera (37)), e.g., millimeter wave camera or infrared camera or detector for the receiver (3) Reflected maser radiation. The tracking of the electromagnetic beam to the receiver (3) Is computer controlled (PC (38)). To avoid unintended side effects of the electromagnetic emission shielded electronics (Abschrimung is (39)). To improve the coverage, the building can be found e.g. on a mountain. The range of the transmission of thought and observation (partly by building (building a city (40)) And forests (tree (41)) Pass) is for example 10 m - 200 km (42).

- [0036]

9 Telepathy of a manned aircraft, an unmanned drone or a helicopter (43) To a receiver (3) By means of the specially modulated beam (4) Of a maser (5). The range of telepathy is, for example 100 m - 20 km (44).

- [0037]

10 Thought transmission from a satellite (45) To receivers on the ground (46) By means of the beam (4) Of a maser (5). The MASER very small Strahlöffungswinkel is fed by a buffered strong energy source, such as a combination of battery and battery atom. For the reduction of the beam diameter and self-focusing effects of the maser radiation are used. The range of telepathy is, for example 300 km - 800 km (47).

- [0038]

11 Telepathy on some 100 major recipient (3) In the event of a disaster using a specially modulated electromagnetic beam (4). For better detection and adjustment of the electromagnetic beam (4) Bear the receiver (3) An electronic label. The transmission is based on the principle of multiplex quasi-simultaneously by rapid switching of the three phased array (30) With 5000 W of average output power. The range of telepathy (partly by building (40) Through) is e.g. 50 m - 20 km (48).

- [0039]

12 Thought transfer to recipients (3) In an emergency using a modified electromagnetic gun (rifle with telescopic (49)) To monitor and anesthesia of receivers (3) Through the walls of a building (a wall building (50)) Throughout. The gun is modified so that it can transmit thought with low electromagnetic radiation power and listen through walls (for example, detection of the change in lung volume).

Classifications

International Classification	A61M21/00
Cooperative Classification	A61M21/00, A61M2021/0055
European Classification	A61M21/00

Legal Events

Date	Code	Event	Description
Sep 8, 2005	8139	Disposal/non-payment of the annual fee	
Jul 1, 2004	8122	Nonbinding interest in granting licenses declared	

Many who have studied the advancement of this technology, extensively such as Cheryl Welsh, mindjustice.org, who has been reporting of brain manipulation through her detailed research for years reporting credible use of mind control technological advancements as seen here which is also an online eBook:

Renee Pittman

The 1950s Secret Discovery of...
The Code of The Brain
~ U.S. and Soviet Scientists Have Developed the Key to Consciousness for Military Purposes ~

Cheryl Welsh, May 1998 - Mind Justice Website

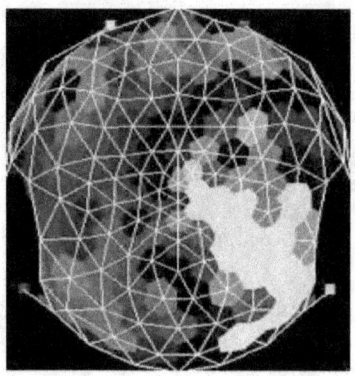

How the U.S. Government Won the Arms Race to Control Man.

A Documentary with Quotes by Leading Scientists, Professionals and Several Independent Sources.

Editor's note:

Some chapters are long, over 20 pages and include the original material because of the importance for education and documentation. The main points of this book are highlighted and arranged with headings so that it may be skimmed.

This is a free online eBook which can accessed at the mindjustice.org website freely, as titled above, and is well worth the read!

The Targeting of Myron May

A Dream Destroyed

CHAPTER EIGHT
To Be or Not to Be That is the Question

The whole idea of the Elite controlled media is to desensitize the masses and reprogram them with the most basic of drives and instincts. This is why sex is heavily promoted and ongoing desensitization to murder and death. The effort is to stop the evolution of consciousness and trap humanity in a state closer to the animal kingdom.

These masterful efforts are capable of stifling the destined awakening, briefly, of higher consciousness and connection of mind, body, spirit and soul. Powerful emotions of kindness, love, compassion and humanitarianism have energy, as does, also, honed energy for evil purposes.

The spiritual battle is very real. Beaten down, destroyed, and broken sadly, Myron May lost the battle rendering him open to programming due to the trauma of his experiences, and devastating and immediate personal losses overnight. The fact is that you are programmable in bitterness, despair, and enmity of injustice. Monarch Programming revealed this using trauma-based dissociation many years in mind control testing objectives. Negative energy is factually the energy which empowers these operations and also those working these programs. These efforts are built around viciously toying with human lives and operate motivated by lower consciousness energy keeping those employed in real delusion as what they are doing is righteous. This mindset is really a psychopathic mentality to deny the horror of "The Program" and justify what is being done as necessary. This is why so many targets feel that there is a demonic connection to targeting due to unspeakable evil coming from operation centers.

Take a moment and imagine that what if everything that I, and many others, to include Myron May are saying is 100% factual. Think on it for a moment. The possibility is not only disheartening but also shocking isn't it? Is not the very thought of a great evil in full operation today and actually doing these things a harsh, dark, and saddening reality?

Myron May had expired and before his body could be put into the grave, the internet exploded with accusations of the episode at FSU as being nothing more than a hoax. Why the hoax angle and promotion of a hoax overall related to Myron May? I think that this was effectively promoted due to the certified mailings being confiscated immediately leaving the shooting open to scrutiny and thereby allowing resulting assumptions to be created of it being yet another False Flag, purely and not the vicious destruction of a gifted young human being's life, resulting from hate, and more importantly, due to the widespread use of powerful, psychological electronic technology and Directed Energy Weapons. Mental illness also continues to be repeatedly highlighted along with subtle but undeniably a connection of a necessity for gun control laws in Florida.

However, the information uncovered via the Gmail letters, and my accessing Myron May's personal email accounts, both Yahoo and Hotmail of which he provided me with the passwords, clearly revealed that what happened was 100% very, very real.

Admittedly, I initially had my suspicions and doubts also as so much disinformation began being publicized. As I stated, I just did not think that this highly educated young man with so much potential, and to live for, and give, would take the route he did. If he had stood and fought, I believe that he could have turned the table. It was later when I learned that his nickname was "Sensitive Joe," a name given Myron May in college, that an image appeared related to May, could confirm a condition of being completely heartbroken.

The fact is, what I compiled earlier as Myron May's Timeline is a compilation of information taken for Myron May's personal email accounts which I copied and pasted not wanting to forward anything to my email account due to the possibly of someone having the much-needed excuse to come to my door saying I was tampering with evidence although he had given the information to me in the Gmail. As I also stated earlier, the Commu Nicate account was closed during the time that I was making several calls to Google for verification of this as being an account listed to belong to Myron May. During that time, in the mist of three phone calls, as stated, the Gmail account was closed down due to ongoing operation center monitoring of my every move, just as his Facebook account was taken down when I tried to save our Facebook interactions on the spot. However, Commu Nicate was verified as belonging to Myron May. Naturally I had to be sure that I was not being fed wrong information.

Everything Myron May detailed escalated to a high-level targeting is classic targeting tactics and designed for an immediate take down effort. His pain, disappoint, great sadness, literally jumps of the page when I first read through all of the material. In fact, the information that May had actually reached out to the targeted individual community as early as November 3, 2014, by emails to FFCHS, gleaned via his email account by emails he and the Director sent back and forth revealed him wanting to take positive action.

This highly educated attorney had hoped to detail what had destroyed this life, and someone knowing that it could be harmful to the truth of what had happened to him, possibly, decided it was time to expire his life and confiscate any evidence or trace. The only reason I believe I ended up with the information is because I was supposed to have it. I would have listed the names involved, in his "being a target" letter detailing his plight, but this is not my battle and I have my own war. I can, however contribute to tell his story and hop to leave a clear understanding of what is technologically, factually, covertly, happening today to many and not just Myron May.

In a climate of global PSY OPS, some believe False Flags are real and can be proven for example by Sandy Hook. However, could it be also as a result of this awareness and awakening of the public today, who no longer accept what is presented as fact, in the controlled media, that someone determined that a different approach had to been taken to regain credibility? It appears so in May's case. The bottom line folks in "The Program" and objective of the program can be viewed as an effort to factually enact gun control laws and any other useful effort compatible to enforcing this and other goals. Most everyone has figured this out. If you don't believe that there is an NWO agenda today, you are in a deep sleep fueled by outrageous distractions. And in recognition of the public's awakening to many levels of deceit today, it appears that a technologically weakened and broken spirit resulted in a mind-controlled patsy.

Could it be that the agency factually pulling the strings thought they could and would get away with destroying a brilliant young African American male's career and life and then stop his effort to tell the world, help other targeted individuals legally, as he hoped, to cement the impression of the desperate need for gun control laws in Florida and to further discredit targeted individuals overall?

The fact is there appears to be a definite connection to current and raging debates today related to gun control laws specifically focused on Florida's Open Carry law. Throw in an added benefit designed to historically continue the depiction of black men as homicidal maniacs, or not human at all, no matter how educated and as historically mentally ill. Then add a decisive effort to create a lasting impression that anyone proclaiming to be targeted individuals is a mentally disturbed conspiracy theorist, and also a dangerous psychotic. Add also if, you will, an ongoing taken down via a focused connection specifically around me and other prominent targeted individuals, and this effort could have possibly been a wrap and all in a day's work. This would have flown had I not uncovered the Gmail Myron May sent to me.

The essential core myth is that everyone is being told from the very beginning that we are to stay stupid and if attempting to smarten yourself up, you will be punished if you go against the grain. And, this applies today to everyone in the form of intentional dumbing down of Americans by the real power structure determined to keep everyone unevolved.

At first when I watched news reports on Myron May, I was struck by an image which flashed showing him in graduation attired, smiling and showing what many individuals know to be factually the Devil horn's hand sign which is being thrown in our faces at every turn as the foundation of the global effort. I was perplexed because today many use this sign to show allegiance. Was not Myron May said to be a lover of bible verses? I later would find a few other images related to this gesture and May.

The Targeting of Myron May

The question is...Are fraternities and Freemasons considered the same?

Freemasonry is a fraternal organization that traces its origins to the local fraternities of stonemasons, which from the end of the fourteenth century regulated the qualifications of masons and their interaction with authorities and clients. The degrees of freemasonry are a gradual system of advancement degree by degree. It retains the three grades of medieval craft guilds, those of Apprentice, journeyman or fellow (now called Fellowcraft), and Master Mason. These are the degrees offered by craft, or blue lodge Freemasonry. There are additional degrees, which vary with locality and jurisdiction, and are now administered by different bodies than the craft degrees.

Lower-level Masons are just dupes being used by the upper-level ones, the so-called "Princes of Masonry." - See more at link below:

In reality Freemasonry is connected to ancient Egypt and the story of Hiram Abiff. Hiram Abiff is the central character of an allegory presented to all candidates during the third degree in Freemasonry. Hiram is presented as the chief architect of King Solomon's Temple, who is murdered in the Temple he designed by three ruffians during an unsuccessful attempt to force him to divulge the Master Masons' secret passwords. The themes of the allegory are the importance of fidelity, and the certainty of death, I might add for demonic force worshippers.

Hiram Abiff of Freemasonry is not an historical character and certainly not a biblical one. Rather, he actually represents Osiris, the Egyptian Sun-god, and the reenactment of the Legend of Hiram Abiff is actually the reenactment of the legend of Isis and Osiris.

The Sign of the Horns

Its origins can be traced to Ancient Greece, and in Ancient Rome it symbolized a curse. The sign of the horns, also corna (Italian for horns, mano cornuta, horned hand fare le corna, to make the horns, or simply the devil horns)

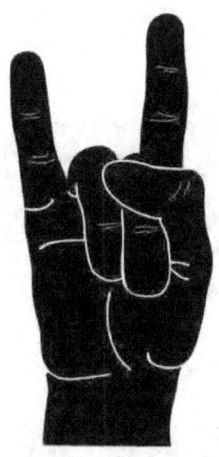

Many today accept this hand sign as satanic. It is also a gesture in Mediterranean countries, often meaning infidelity, the "devil sign", mano cornuta or horned hand, inspired by Anton LaVey. Used by Satanist, in the heavy metal music industry, it has turned it into a sigh we use for "rock on" as has its use in certain fraternities. American universities have adopted it as a sign of support for their athletic teams and when displayed with the palm facing inward it can denote the Phi Beta Sigma Fraternity, Inc., of which May was a member. The Kappa Alpha Psi Fraternity uses the 666 hand sign which in reality is the three sixes sign of the Illuminati.

Alpha Kappa Alpha Hand Sign Nupe kappa alpha psi

Delta Sigma Theta

Three 6's Illuminati Hand Sign

The fact is everybody wants to belong to something. It is human nature to join and want to be a part of a larger movement. The Fraternity and Sorority system is no different than the Masons and the Eastern Stars.

The Targeting of Myron May

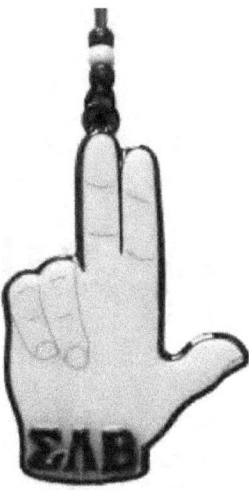

Both top and bottom are Pi Phi Hand Signs

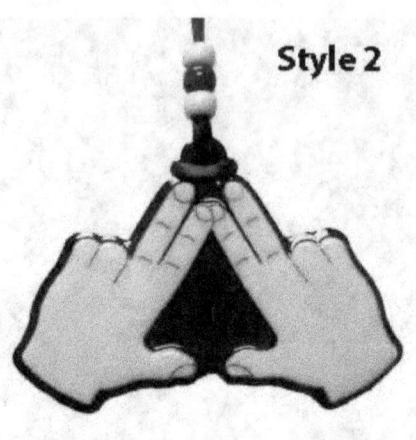

Skull and Bones

The Skull and Bones are described as a group of upperclassmen at Yale University who take part in selling their souls to Satan to get into the upper echelon of society.

Black 'Skull and Bones / the "Boule"

"The first Black Greek fraternities and sororities were created in America in 1904 in Philadelphia, by Dr. Henry Minton and five of his colleagues and culminated in the 1960' s."

The Boule, (an acronym for Sigma Pi Phi) and pronounced "boo-lay"), was formed to bring together a select group of educated Black men and women who would enjoy many perks of which a majority of disenfranchised blacks were not privy too.

The Boule recruits top Blacks in American Society into its ranks. Today, 5000+ Archons, (male Boule members) and their wives, (Archousais), with 112 chapters, make up the wealthiest group of Black men and women on the planet. "Archon" means "demon" - the kind that likes to stay hidden.

The question then became, to whom do the Yale University Skull and Bones, and The Boule really serve? Many would respond, the satanic global elite! And, as long as members conform to the rules, the riches will be in abundance; if not, down comes the hatchet on many levels

Historically, with the advancement of highly educated blacks, there also became a necessity to indoctrinate them for control and usefulness. These affluent groups were then sworn to uphold, and have a vested interest in upholding, and protecting the ideals of the Elite System and its advancements.

Without a doubt, the fact is today that anyone in any powerful position, to include, even a President, judges, politicians, directors of powerful government agencies, including Black lawyers, doctors, engineers and accountants, etc., are members of this secret club at various levels.

One thing cannot be denied, anyone who acquires any degree of success after connecting with organizations which have historic development based on sinister doctrine has in some way aligned them, by oath, with this doctrine. This is especially exampled by the entertainment industry in the form of Luciferian symbolism displayed entertainment superstars. And many distractions continue to trick the mind of the population away from this reality through glamorization.

Whether through music or films or any other industry the ruling elite use these recurring symbols to show their influences. As with other industries, it represents their signatory; their way of telling us how they control and influence the world, whether we know it or not. They also believe that these occult symbols have a subliminal unconscious effect on people:

Many people particularly in the alpha state, are blindly unaware of their influences and have accepted these symbols. Some consider them to be cool and even hip, without understanding its low vibration effect. A great example is shown if you Google entertainment industry hand signs of the Illuminati. The "All Seeing Eye depicted as covering of one eye is a favorite among many images promoted among celebrities connected to subliminal influence and NWO ideation.

Spiritual warfare is what Myron May described in his own words, when he stated being in intense conflict against self. Today, it is obvious that he lost the battle. From a Christian perspective, spiritual warfare is the mental battle that takes place when you want to do well and obey Jesus, but you find there are things pulling you in the opposite direction. However, in this regard, it is pivotal to understand that Myron May was likely placed on a "target list" by possibly someone he worked for, prosecuted, or did business with professionally. One thing is certain, this technology; due to its powerful influencing capability can also put the targeted individual into spiritual warfare too. This is especially true if requests are beamed at the target via radio frequencies to go against the target's will and personal belief system and to cause harm. Repetition, a decisive practice in these operations plays a powerful role also in the programming process.

Myron May was highly education and had expertise in a lot of areas and most understand that attorneys are frequently bought into the arena of shady individuals officially and unofficially. May practiced:

Administrative and Public, Business, Consumer, Criminal, Insurance, Labor-Employment, Litigations: Commercial Litigation:

The Targeting of Myron May

Personal Injury, Other, LGBT Law, Workers 'Compensation, Appellate: Civil, Appellate: Criminal, Disability Law.

I had to ask when I kept seeing Myron May flashing this hand sign, does May and many others know exactly what they were showing allegiance too in organizations which portray themselves as community humanitarians? And, if awakening to the reality of actual programming and pledging allegiance to a doctrine founded on satanic principles, even if albeit naively, would this put a religious person into conflict and weaken the person's perception of God on a spiritual level?

What is happening today is in direct opposition to the principle of unconditional love for all life, and global unity, and is against the unification of humanity as a whole. A unified humanity cannot be effectively, collectively controlled by evil when holding a higher consciousness. However, if divided through racism and hatred, or by perceptions based on lack and limitation humanity can be controlled and human beings can be turned against each other and continue to be used. One thing is certain; the dynamic of electromagnetic energy draws positive or negative energy to you and creates our reality via our thoughts and emotions. This is the nature of the Law of Attraction.

Blackmail is always part of the deal when selling of the human soul. May reported that he was told that if he did something horrible, that he would be rewarded with freedom and his life returned to him. He was deceived.

The Black Greek fraternities and sororities by their very existence are factually deceptions disguising the practicing of a corrupt shadow of a stolen doctrine remade into a carbon copy of a doctrine taken from the ancient African Mysteries of Kemit. Kemit as we know was the original name of ancient Egypt before it was later named Egypt by the Greeks. In reality there are no Black-Greeks.

Today there are Nine Fraternities and Sororities that make up this order and are called the Divine Nine. The people who started the fraternities and sororities were probably unaware or unwilling to see the 'Stolen Legacy' of this Greek system used to deny contributions of

4,000 years of African religion and culture and blind civilization to reality. After, most do not want to admit after all that Egypt is located in Africa.

The Greeks adopted this system from Egypt, modified it, and then repackaged. Years later it would be sold back to the children whose heritage from which it was taken. African American fraternities and sororities successfully accomplish this in naivety it seems, or do they?

Today, the signs of original Egyptian doctrine have been twisted negatively in support of a Luciferian agenda and symbolism is the key. This is by abuse of universal principles and the electromagnetic spectrum. The fact is the Egyptian Mystery School represented unification and protection of humanity. The Greek version represented separation and deception founded on a stolen legacy.

In the minds of unconscious men and women who make up the so-called talented tenth of African-American society, belies the thread of the Hellenistic society. This Society has, as the conqueror, changed the truth into a lie and formed the children of the founders of the Mystery system into the slaves of a false reality based in the theft of that powerful educational system founded on intentionally fraudulent representation of history. A person is nothing without their history.

Today, the reality is, that a small group of international elites (known as the Ashmedai's 2069, 1001 Club, Bilderberg Group, Bohemian Grove, Committee of 300, Council on Foreign Relations, Freemasonry, The Illuminati, Le Cercle, Majestic 12, Men In Black, Nine Unknown Men, Pilgrims Society, Royal Institute of International Affairs, Skull & Bones, etc., Trilateral Commission, or more are broadly viewed as the engine enforcing the New World Order and of which controls and manipulates governments, industry, the financial system, banks, and media organizations worldwide who are unequivocally literally Hell bent on a one world governance. In this reality any and everyone can be used.

In a powerful effort today based on powerful, technological, manipulative control at the highest level in the pyramid, many do not see or understand the aggression and dedication to the master plan. Most cannot fathom that the average man or woman can and are being perceived as nothing more than mere Sheeple.

The main foundation for the FSU shooting as a hoax is based on law enforcement's response's time and the whole episode being wrapped up in approximately three minutes to include May's shooting spree and the dramatic conclusion as reported in the media. To many this made absolutely no sense at all and set the stage for a hoax. The three-minute publicized total ordeal, is completely illogical and now appears to be an intentional deception to confuse the public and make some question the reality of the shooting overall.

A November 21, 2014 press statement from Tallahassee Police Public Affairs Officer David Northway to select news media, some would argue, appears to imply a media connection and reads:

"The Tallahassee Police Department would like to thank our media partners for their patience with the incident, in particular releasing the identity of the victims. This has allowed TPD to conduct interviews with some of them and begin to piece the facts of the case together."

If not a hoax, many believe that law enforcement had to be, at the very least, warned in advance of May's coming which would then also reveal him as being used. Thousands of discredited targeted individuals know that when targeted, it is a 24 hour a day, 7 day a week effort for months and years without fail. But because no one believes in the technology's uses through well-crafted media promotions, disinformation of the truth can be suppressed.

A person who witnessed the shooting reported what happened from start to finish. He posted what appears to be a more credible version of the shooting and its timeframe.

Robert: The FSU police did an amazing job in their response time. They were there within 10 minutes of the first shot. The actual shooting lasted about 5 minutes. After firing shots at about 5 people, he walked into the front lobby of strozier. He stood there for about 4 or 5 minutes and messed around on his phone, then he walked outside. He then waited around the front steps of strozier for no more than three minutes. Police arrived at this point. They surrounded him and he stood still, with hands half raised. After about 10 seconds of yelling "put the gun down", he hadn't moved an inch. It was then that they all proceeded to fire multiple shots at him. I know this because I was there in case some question the validity of the source.
1 hr · Unlike · 👍 4

 Christine: Wait so you watched this guy shoot his gun and you stuck around?!
1 hr · Like

 Robert: it all happened really quickly. I heard the first shot and didnt know what it was at first. I looked around and didn't see anyone running and didn't hear any people making noises so i didn't think anything of it. Then a minute or so went by and i see some guy walking and then as he passes by this kid he pulled out a gun and shot him. Then I realized what was going on. With all of the shock and adrenaline it takes a second to realize what's going on. I was standing on the benches under the trees. The kid he shot at was sitting at the tables over closer to dorman. By the time all of this hit me, he was about 25 feet away. He fired one shot at a friend I was with and one at me. The bullet he fired at me missed, but the other bullet hit my friend and he's one of the three in the hospital.
1 hr · Like · 👍 4

 Christine: Wow. So great to hear from a witness. I bet the police interrogated you for hours. I was in Strozier for the whole thing but didn't even hear the gunshots.
1 hr · Edited · Like

 Robert: you'd be surprised. I waited around for 30 minutes after to try and give my testimony but they ignored me and told me to go away.
1 hr · Like

Eyewitness Account of Student

If this is factual, the portrayal of the incident as taking just three minutes could be then considered PSY OPS with the desired result to again, confuse and distract, but leave an impression of key issues, again of targeted individuals, and conspiracy theorist, as factually mentally ill, and that the public should not believe them and that gun control laws must be passed immediately. Again, these are well thought out operations especially in relation to efforts for gun control laws and efforts to keep secret technology use covered up.

Below, in a hopeful effort not to offend anyone, I have also included the crime scene photos. The body language in the second image of the man standing to the right holding the flashlight is telling by his body language to me.

The determination of the whole thing as being a hoax abounds in spite of these images. If this is a hoax then it is a masterful dupe of America and Americans as a whole literally down to the letter!

Understand that of two groups heavily under full attack today through massive discrediting efforts, again are conspiracy theorist and targeted individuals, and also both via the useful tool and partnership of the field of psychology. In fact, psychologist report the origin of conspiracy theorist includes projection; and as being the personal need to explain "a significant event (with) a significant cause;" and the product of various kinds of states of thought disorder, such as paranoid disposition, ranging in severity to diagnosable mental illnesses of which if they could they would likely medicated millions with anti-psychotic medication. In reality this could also be said also about psychologist who needs to be medicated as they play their role in what is happening today.

Conspiracy theories are nothing new and many have learned that they typically have proven many realities. A good example of a conspiracy theory as being fact base is related to the Pearl Harbor attack.

In 1945, various parties high in the U.S. and British governments knew of the attack on Pearl Harbor in advance and may even have let it happen or encouraged it in order to force America into war via the "back door." Today, many report the same dynamic for war today, via the Twin Towers and also for gun control?

Excerpt from the Liberty Voice article below entitled Nazism and Psychiatry, by Lucille Femine on September 16, 2013, reveals an ongoing connection of world control ideals of Nazi ideation.

Saved under "Political Right" in the Liberty Voice

The Nazi era is not only the most devastating example of the complete suppression of human rights; it was quite intimately tied to psychiatry. In actual fact, psychiatry was the most powerful and driving influence behind Nazism.

It is commonly believed that Hitler ordered the mass execution, including cruel human experimentation, of Jews, gypsies and homosexuals during World War ll. Although just as responsible, he did not order that dark turning point in history, he simply signed his consent.

It was actually psychiatrists who initiated and carried it all out on such a massive scale. Nazis were the first in history to use extermination camps and all done with such chilling organization.

The two main areas that psychiatrists concentrated on were sterilization and euthanasia. They were responsible for reporting these "patients" over to the authorities and from there, to the gas chambers, killing over 200,000 people deemed mentally ill. These included many thousands of feeble-minded children. The real intention was to rid the master race of "undesirables."

Psychiatry expanded their vast control by broadening the definition of mental illness to include political disobedience. Thus, psychiatry became a tool of and ally to the government, particularly the Nazi-controlled regime.

Once a person was labeled mentally ill due to some involvement in a political disturbance or disagreement, he could be sent away to a mental institution indefinitely and, of course, without a trial. What better way to squash the right to defend oneself or reveal some injustice, especially since one is now "crazy"?

Read more at http://guardianlv.com/2013/09/nazism-and-psychiatry/#carDVBgctHx1tYGl.99

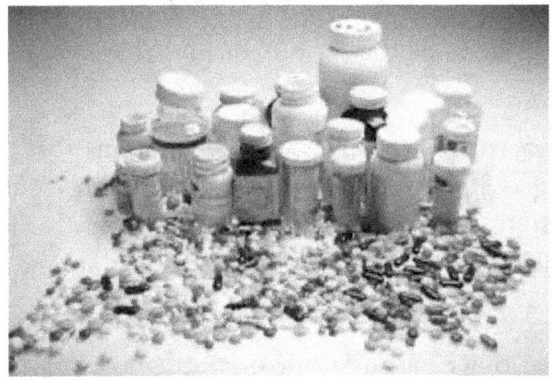

Add biometric radio frequency technology

Perhaps one of the most powerful revelations that rendered the May shooting as a possible hoax was a report of an "Active Shooter" drill reported to have happened specifically on the Florida State University campus approximately two weeks prior to the shooting.

If true this too would warrant credible consideration for a hoax. But in light of other media falsified report, and again media partner connections, and the logical revelations that it took at least 10 minutes or possibly slightly more for cops to arrive, the media could now be revealed as a true co-conspirator.

In fact, this theory erupted because of what appears to be a televised interview by FSU Police Chief Perry who states during a news conference, "officers had a good memory marker of how to respond to "Active Shooters" due to a recent drill, and as a result, knew what was appropriate for this type of situation".

The six officers involved, the media reported later reported were placed on administrative leave pending the investigation. Later they were exonerated by Grand Jury as taking appropriate action.

This was reported by GTN News out of Gainesville, Florida

There were also another televised news reports, reporting an "Active Shooter" drill on or about November 14, 2014, happening in Winter Haven, Florida as an unannounced drill at Jewett Middle Academy in the Polk school drill:

Nowhere in mainstream media is the GTN news report of an actual active shooter drill happening on the FSU campus. However, the above drills were also articles on washingtonpost.com and huffingtonpost.com, etc.

Excerpt:

"A school lockdown drill in Winter Haven, Florida, angered parents whose children apparently texted in alarm, with one noting that a police officer with an assault rifle came into a classroom.

Initially, only three people knew the lockdown at Jewett Middle Academy on Thursday was only a drill, police said. Parents, teachers and students were not told in advance about the scheduled event because, the school district said, the element of surprise is vital to make the drill effective..."

I have heard it said, that perhaps, Myron May's death was certain but that those whom he supposedly injured, with one separate individual reporting that a book saved his life by stopping the bullet also added to the suspicion. However, we know today that there is no law against crime scene manipulation or bogus news reports in controlled media reports. The fact is, again, what May wrote unequivocally, 100% proves him as a targeted individual to the letter, literally.

A crisis actor, named Blair Stokes whose tweets and testimony was featured all over the media, combined with an odd and awkward televised interview, during which she appeared smiling and delightful, was later revealed to work in the media as a crisis actress. This Stokes is said to be employed by a firm which specializes in "Crisis Communication" and "Behavior Change" designed to "Change Public Opinion" on a specific issue in its website.

The CNN interview is dated November 23, 2014 during the day, with Stokes, and noteworthy, three days after the shooting.

The fact is, if May was being used and tracked as a patsy would be allowed to hurt Florida State University students? However? In the selfish objectives' necessity for gun control laws anything is possible as thousands die in war and along with May's life possibly becoming useful for the sacrificed for a cause.

Later in this chapter, you will also see yet another similar Facebook post by another person labelled as a mass shooter named Amanda Miller. The Jared and Amanda Miller shooting spree took place in Las Vegas, Nevada. Amanda Miller made a similar post, similar to Myron

The Targeting of Myron May

May's as an omen on Facebook too. May as shown earlier posted and mentions that he has a handler. He also mentions a handler; perhaps in his operation center likely nudged him to even posting this on Facebook strategically.

Has anyone here every been encouraged by your handler to kill with the promise of freedom?

I have also used the blow article previously to substantiate the fact that Directed-energy weapons are on the horizon as published by the article below in the Houston Chronicle in 2005 for those who have not seen it before.

Does this article help to confirm that Directed Energy Weapon usage is very real and in full use today, albeit covertly? May reported repeated hits in his chest relentless and escalating.

Houston Chronicle 2005 via Associated Press Reporter

There were three red flags for me with May. One was his volunteering to help others with legal help in California and brushing me off after my offering to pay to get him certified in California to help. The second was the unusually long breaks when messaging each other by him. This led me to believe he was possibly being coached. And the third as shown was the unmarked postal receipts which I felt typical of certified mailings which are substantiated by at least date and time stamping by the USPS green and white receipt.

One thing is certain, the manner, in which the person identified as an FBI agent acted while in my home by oddly positioning himself behind me as I sat at the table talking with the Postal Inspector, part of the effort was a definite act to intimidate and also a hope to frighten me.

One of the most revealing cases of a possible connection to gun control laws is asked, "Why would May choose the FSU campus?" Mr. Taunton reported, during his interview that May loved Florida State University and was a very proud Alumnus.

Some would argue that efforts concerning "Open Carry" Gun Control are 100% part of the NWO agenda of which mass shootings are playing a major role in hopes to change the public's perceptions.

The Fact is there are several colleges who are in opposition to the SB 234 Bill in Florida and are on the record in opposition to the Bill. This is as the battle rages for US citizens guaranteed a Constitutional right to carry.

Here is an example below:

Florida Open Carry

Welcome to Florida Open Carry!

Open Carry is a grass roots movement made up of people who seek to protect and expand their individual right to keep and bear arms and are willing to exercise that right. Open Carry refers to the act of carrying a firearm in plain sight.

In Florida, you can legally open carry a loaded firearm while engaged in, or going to and from, Fishing, Hunting, and Camping. With some planning and preparation, a law-abiding person can open carry a firearm in public and stay in compliance with the law.

Responsible, law abiding friends associated with Florida Open Carry hold meetings throughout Florida in an effort to raise awareness in the community. Our goal is to help educate others about their right to legally open carry and advocate for expanded open carry as allowed in the vast majority of other states.

If you would like to help expand your right to carry openly in Florida, come out for some of our great fishing and camping events to promote OC!

Senate bill would allow guns to be carried openly.

SEBRING -

Four Legislative bills seem poised to change Florida gun laws.

- SB 234/HB 517: Lets people with concealed gun licenses carry weapons openly. Also allows carrying or storing firearms in a vehicle.

- SB 432/HB 155: Doctors can be fined $500 or more if they ask patients about guns in the home. The original version carried a $5 million penalty.

- SB 402/HB 45: Fines local governments that enact gun laws stricter than the states.

- SB 956/HB 4069: Repeals laws against purchasing rifles and shotguns in contiguous states.

- SB 234/HB 517

- SB 234, sponsored by Criminal Justice chair Sen. Greg Evers, R-Baker, passed his committee, 3-2; and is now in the Judiciary committee. Companion HB 517 passed the House Criminal Justice committee 10-3; it's now in the House Judiciary committee.

If the open-carry bills pass, Sheriff Susan Benton imagines how Circle K or Walmart customers will react when other customers have guns on their hips.

"Do you think that the clerks will calmly ask the citizens to produce their license to carry a firearm, or will they call 911 and feel like they have an emergency?" Benton asked. "Most sheriffs that I have spoken to are against this bill."

Two weeks ago, Florida sheriffs testified in Tallahassee against SB 234/HB 517. They're concerned that deputies responding to a conflict would have to disarm everyone, instead of focusing on the perpetrator.

Polk County Sheriff Grady Judd, an NRA advocate, worried that tourists would be frightened by openly carried guns. "It's just intimidating and threatening."

Evers' original bill allowed college students, employees, or faculty to carry guns. However, on Jan. 9, Amy Cowie, 20, watched her twin sister die after Amy's boyfriend accidently shot Ashley with his assault rifle at a Florida State University frat house.

Sen. John Thrasher, R-St. Augustine, is a friend of the Cowie family. Thrasher is the Rules Committee chairman; he could block bills from the Senate floor. Evers struck an agreement with Thrasher to prohibit guns at college, keep the ban on firearms at K-12 schools, and allow stun guns at colleges.

NOTE: John Thrasher is also the Florida State University President.

"It seemed to be a big deal," Evers told the Miami Herald. "There are still a lot of good things left in the bill." He said the bill would protect people with concealed weapons permits from getting in trouble for accidentally revealing their guns.

Sheriff Benton disagreed: "I don't think that one of our deputies would arrest a person who has a license to carry who inadvertently shows his or her firearm when they take off their jacket while getting into their car."

"I feel that this bill totally lacks common sense," Benton said.

College presidents lined up against the bill.

"The presidents in the Florida College System are on record in opposition to the proposed SB 234. We favor current law. The vote was unanimous," said Norm Stephens, president of South Florida Community College.

State Rep. Denise Grimsley, R-Sebring, supports HB 517: "I'm proud of the work the Florida House has done over the years to preserve and defend our citizens' Second Amendment rights, and I believe we will continue to ensure that law-abiding citizens are able to exercise their right to bear arms in our state.

"Current law allows registered students, employees, or faculty of colleges or universities to carry a stun gun or nonlethal electric weapon or device that does not fire a dart or projectile," she added. "The House bill does nothing to change this."

Marion Hammer, lobbyist for the Unified Sportsmen of Florida and former NRA president, said Evers' bill would not have prevented Cowie's death because the accused shooter is under 21 and therefore isn't eligible for a concealed weapons permit.

"There's a lot of safety by allowing guns on campus," Hammer told the Miami Herald. "That's how a lot of us protect ourselves, because law enforcement can't be there when we need them. Law enforcement is not stopping rapes on campus, and not stopping a lot of crimes."

SB 432/HB 155

There's a gunfight between Second Amendment advocates and doctors, and the pistol Eros may have the law on their side.

Evers sponsored SB 432, which passed his Criminal Justice Committee 4-1. It's still in the Health Regulation Committee. Companion HB 155, by Rep. Jason Brodeur, R-Sanford, passed the House Criminal Justice Committee 9-6, and it's now in House Health and Human Services Committee.

The bills would ban public or private doctors or other medical personnel from asking whether a patient or the patient's family owns firearms or has a gun at home.

Gun lobbyist Hammer told legislative committees on Feb. 22 and March 8 that the medical community - specifically the American Academy of Pediatrics - has been "pushing a gun-ban agenda and have been bringing their gun-ban politics into examining rooms."

A Politi-Fact checked out Hammer's claim on Feb. 22 by searching the AAP website www.healthychildren.org, and found her to be correct: AAP changed a statement which did say, "The most effective way to prevent firearm-related injury to children is to keep guns out of homes and communities. The American Academy of Pediatrics strongly supports gun-control legislation. We believe that handguns, deadly air guns and assault weapons should be banned."

AAP's statement has softened: "The safest home for a child is a home without a gun. However, if a gun is present in the home of a child, the gun should be stored unloaded and locked, with the ammunition locked separately, so that a child cannot access the gun."

SB 402/HB 45

Sen. Joe Negron R-Palm City, and Evers sponsored SB 402; HB 45 is carried by Matt Gaetz R-Fort Walton Beach, the son of Sen. Don Gaetz. HB 45 passed Criminal Justice 10-4, and proceeded to Community & Military Affairs. SB 402 passed Criminal Justice, 3-2, and Community Affairs, 9-0.

It would void, for instance, zoning ordinances designed to restrict or prohibit the sale or manufacture of firearms or ammunition. It would also eliminate provisions authorizing counties to adopt an ordinance requiring a waiting period between the purchase and delivery of a handgun.

SB 956/HB 4069

Sen. Alan Hays, R-Winter Garden, sponsored SB 956. It was referred to Evers' Criminal Justice committee. HB 4069 by Rep. Jose Felix Diaz, R-Miami, passed House Criminal Justice, 13-0, and Business & Consumer Affairs, 10.0

It makes it easier for Florida residents to buy rifles and shotguns in contiguous states. It eliminates the authority of counties to adopt ordinances requiring a waiting period between the purchase and delivery of a handgun, and provides that public funds may not be used to defend the "unlawful conduct" of local officials who oppose the law.

See more at:

http://highlandstoday.com/news/agri-leader/2011/mar/25/LANEWSO1-senate-bill-would-allow-guns-to-be-carrie-ar-305488/#sthash.sv8im547.dpuf

Could the effort have been decided to be evolved by some covert psychopath, and kill two birds with one stone by attacking Christianity as a false religion of homicidal maniacs while ongoing promotions of Luciferian ideals too? The fact is, these efforts when connected to the name Targeted Individual continue the perception of targeted individuals as mentally ill and that the technology testing program is a delusion of thousands of people who are factually mentally ill disturbed as proven by Myron May along with the continued perception of black men, no matter how educated as murderers, who are in reality used as a patsy to assist in global genocide.

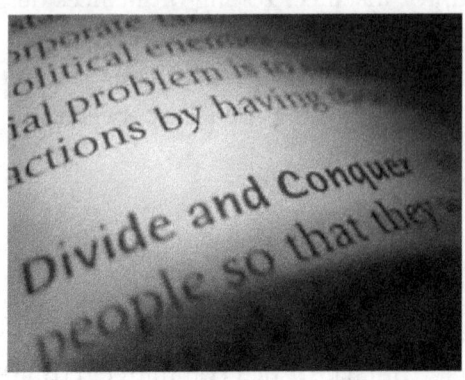

Folks, African Americans still remain just 13% of the population. In this regard, all of the puppets being used by spreading hate, both black and white, and other, when this is considered, reveals that the odds are stacked against African Americans inciting a race war. However, this is not impossible with algorithm-based radio frequency influencing prefaced by media illusion management of an actual threat of 13% of the U.S. population as factually being a threat. Today, individuals, groups and large population manipulation is a technological reality, and must be covert, in order to be effective.

CHAPTER NINE
NWO Luciferian Agenda and Strategic Distractions

Ferguson, Missouri
Myron May
Bill Cosby

Is there a connection? One thing is certain, ALL lives appear to be expendable today.

(Excerpts from mindcontrolblackassassins.com)

"Most people have come to associate Bill Cosby with intellect, civil responsibility, clean old fashion humor, and childlike affection.

Cosby is a world-renowned comedian, actor, author, television producer, educator, musician, humanitarian, and activist. Cosby is a living legend, trailblazer in Black Entertainment, multiple Emmy and Grammy winner. He has been awarded numerous honorary degrees from Yale University to Sisseton Wahpeton College on the Lake Traverse Reservation in South Dakota.

On October 27, 2009, Cosby was presented with the 12th annual Mark Twain Prize for American Humor. In a British 2005 poll to find The Comedian's Comedian, he was voted among the top 50 comedy acts ever by fellow comedians and comedy insiders. He received

Kennedy Center Honors in 1998. He was awarded the Presidential Medal of Freedom in 2002 for his contributions to television. He won the 2003 Bob Hope Humanitarian Award.

Bill Cosby that you have become accustomed to is an illusion. There is another secret side of Cosby, William Henry Cosby, Jr., that you won't like.

MK ULTRA & Jell-O Man, Don't Let Him Near Your Daughters!

Bill Cosby, the Jell-O man, was enlisted as a hospital corpsman in the U.S. Navy from 1956 to 1961 during Operation Paperclip and CIA/MK ULTRA. He had been assigned to extremely sensitive and secret notorious Operation Paperclip, CIA/ARTICHOKE-MK ULTRA-MONARCH and U.S. Biological/Genetic warfare sites, among others, the Naval Medical Center Bethesda in Maryland, Marine Corps Base Quantico, Virginia, U.S. Naval bases in Newfoundland and Argentina.

Needless to say, Argentina under the influence of totalitarian and fascist Juan and Eva Peron was a safe haven for the CIA Operation PAPERCLIP's most notorious war criminal Nazis and SS like Hauptsturmfuhrers SS Dr. Josef Mengele, the Demon of Death and Klaus Barbie, the Butcher of Lyon.

In 1951, Erich Traub, Heinrich Himmler's biological warfare expert, was secretly harbored at U.S. Naval Medical Center in Bethesda. Alfred Hellman was also working out of the Naval Medical Center in Bethesda developing and experimenting with (Traub) `biological warfare oncogenic (AIDS) viruses for the Navy.

From 1951, the Naval Medical Center in Bethesda had also been the site of the CIA's Project Bluebird-Artichoke Assassination programs, former Korean War POWs were subjected to various behavioral modification programs, including the use of experimental drugs, special interrogation methods; and hypnotism all for what the CIA deemed "offensive objectives."

From 1953, the Naval Medical Center in Bethesda had additional been the site of the secret CIA/MK ULTRA mind control programs using mind controlling drugs on servicemen, and programming mind-controlled slaves for Project MONARCH.

Reportedly, Cosby worked with an extremely sensitive class of soldiers that had been involved in the Korean Conflict (1950-53), which may have involved Korean/Chinese Brainwashing Techniques, the LIDA Machine, CIA Projects Bluebird-Artichoke, and Frank Sinatra's type of Manchurian Candidates.

Whatever Cosby learned from the U.S. Navy and MK ULTRA, he applied unmercifully for whatever covert purposes in the mega movie-television industry. The Jell-O man is accused of being a hush-hush psycho-narco mind bender and serial rapist.

From 2004 to 2006, at least 13 women have come forward to accuse Jell-O man of covert narco drugging and sexual assault. These women are said to currently reside in California, New Mexico, Florida, Nevada, Colorado, Ohio, and Arizona. Of nine "Jane Doe witnesses," scheduled to testify against Jell-O man in a civil sexual assault case, six were represented by Pennsylvania lawyer Joyce Dale, executive director of Delaware County Women against Rape.

Two other women, according to civil filing, are believed to be seeking to retain counsel to pursue civil claims against Jell-O man. In addition to the other women claims, civil lawyers also sought New York police and prosecution records relating to a 2000 abuse claim reportedly lodged against Jell-O man by a "Cosby Show" cast member named Lachele Covington. Cosby has denied all abuse allegations, none of which have resulted in criminal charges of rich, powerful and famous Luciferians with the power of the Empire State of Mind.

What else does Anton LeVay, Sammy Davis, Jr., Bill Cosby and Robert De Niro all have in common? They all have played the Devil or Lucifer in major Hollywood television and film productions."

Is there a connection on three stories running simultaneously, Ferguson – Martial Law, Myron May – Gun Control and impression that mentally disturbed individuals are also accessing weapons, and Bill Cosby – distraction and programmed fall guy?

Don't be misled, the battle between Good and Evil is an ancient conflict going back tens of thousands of years. The languages and names have changed but the game is the same. For hundreds of years, people have willingly given up salvation to serve Satan and reign in Hell which today appears to be an Earthly endeavor.

There is a saying that says, God give you talent but Satan gives you fame. If anyone reaches any success in life and bases that success on faith in God, the individuals can become a threat to the NWO due to the ability to reach millions via the media. Either a man who has reached success must do the work of Satan or be reduced to a condition akin to that of slavery. He will no longer be an inspiration to anyone to struggle against universal racism, and the principalities of Evil or the NWO. In fact, once a target of Luciferians agendas, very few people understand the vital codes to overcome a comprehensive offensive program designed to destroy them by legions of networks of the Secret Satanic Brotherhood.

Myron May's initial, unbroken connection to the Christian faith doctrine, and practices of Christianity is not necessarily as biblically revealed as the Will of God. What is mean is that within the African American culture the focus is on "Faith in the Will of God" in the ongoing African Struggle for justice, freedom, and liberation in America.

Christianity for Black people has become almost equivalent to the Rod, Sword, and Shield to overcome by the Will of God – the constant struggle against racism, and the principalities of evil. Again, May loss this struggle. However, understand today we are literally dealing with evil high places, such as space-based as defined by the Dennis Kucinich, Space Preservation Act of 2001.

As the Luciferian agenda marches on empowered by a satanic beat the formulation of the one world worship of the Beast, a perfect scheme. Without a doubt there is a hope to destroy religious images, practices, and faith in religion such as Christianity, etc., and using someone known to be a devout Christian is good strategy.

Additionally, the fact is the South, where Myron May ultimately met his demise, always has had cultural conditions which actually date back decades are part of the fabric of America. This is in consideration that the type of targeting which May was being programmed for could have originated from, for example, Arlington, Virginia or a military base in Florida with local police foot soldier doing their part in an ongoing destructive stalking campaign:

Excerpt, Dr. Jason Smith, PhD., Anthropology:

"There is a universal basal imprinting of all people, regardless of class, in a given sociocultural "epoch" in the ideological template of that Epoch or Period."

"People are not born with a blank slate for long. Imprinting begins with birth and proceeds rapidly apace so that by the time a child is a few years old it is well set in."

"This mental template or imprinting 'colors' the way people behave in the most profound underlying causal way."

"The mode of production (the relationship between technology and social organization) of any society will determine the type of ideology that can occur (the kind of idea system that is possible.) Within Servitude Epoch societies, the primary mental template will always have two poles as opposite sides on a broad-spectrum in sociocultural evolution, we have a primary mental template also existing as a spectrum with two opposite poles: one (left) side is simple selfishness/self-centeredness, and on the other (right side, outright sadism."

Racism can appear within the mental template of selfishness. Racism or bigotry is a small part of the bigger picture.

I am concentrating on these theories as it might apply to the social or cultural conditions in the state of Florida. Here, in Florida, I believe the strong template of selfishness and sadism and also racism has been entrenched since at least the days Confederacy and the Civil War. It was perhaps attenuated or amplified from the slavery days of the South. I say this because, before the Civil War, the territory of Florida was the scene of a serious of violent and bloody conflicts known as the Seminole Wars, where sadism and racism was the norm of the day rather than the exception. Note that the Seminole Wars were also conducted against escaped slaves. Because slaves who could reach Spanish Florida were essentially free. Some intermarried with the Indians and formed tribes of Black Seminoles.

Myron May's ex-girlfriend reported that when she met him in the parking lot, the sight of him jarred her. Gone was the well-groomed, meticulously dressed man who had taken her on dates for most of the past year 15 months.

He was gaunt, haggard, disheveled and wild-eyed reduced to nothing as someone technologically intended to be his fate. He wore a borrowed T-shirt and a pair of too-small running shorts. He was barefoot. He had thrown away his shoes, he told her, because he was sure they were bugged by the cops who were following him. Factually, it is 100% likely that May was being tracked biometrically via satellite real time, microwave artificial technological harassment, and energy weapon drone attacks.

May begged her for help. In his desperate hope to escape the ongoing technological harassment and network stalking, he told her he needed her to rent him a car, so he could slip out of town unnoticed. At this point it is clear that Myron May did not understand the brevity of these operations nor a biometric tracking capability which can track a person biometrically any place of the face of the earth by, for example, Eagle Eye.

This was Oct. 8, and as reported, the night before his ex-had made an official police report detailing May's crumbling demise. This also was just a mere six weeks before May would walk into the Strozier Library on the Florida State University campus.

In interviews with the Tampa Bay Times on Friday, November 21, 2014, May's friends described their frustrations through efforts to get help for their friend.

"You have to commit a crime to get the help you need. Why isn't it the reverse?" said Name Deleted a Houston lawyer who described May as one of her best friends. "This could have been avoided, the entire thing" she told the media.

Or could it have been avoided if May was factually technologically targeted which, again, his documents indicate that he was and heavily? The answer is a resounding NO!!!"

The fact is that there is no hospital cure for technological psychophysical harassment from highly advanced technological operation's today. And, being committing to a hospital and accepting medication actually helps to seal the target's fate as May sadly must have realized in hopeless despair by agreement to do so.

Does this make in any way what May did right? Absolutely not!!! As stated, however, again, if technologically harassed, nudge, manipulated, influence, or tortured into heinous action, those at the helm of this technology are even more responsible and May is without blemish in the eyes of God.

But do the minions or their real employers really care? With such evil living in the hearts of some men, in reality this Earthly experience is the last stop for them and they know it. This is the scientific foundation for trans-humanism. This was one of the focuses of the 2013 Bilderberg meeting. The statue placed outside of the location of the meeting was yet another, blatant sign of yet another satanic agenda. This hope since creation by very real evil forces has been ongoing as efforts to change what God created and a hope today in the trans human agenda. The image is intentionally depicted as birth from water as a symbol of amniotic fluid in the womb most likely.

NOTE: Transformers: Age of Extinction is a 2014 science fiction action film based on the Transformers franchise.

The Targeting of Myron May

Myron May stated that he resigned his position with the DA's office in New Mexico because the technological harassment was extreme and relentless verbal degradation played out technologically inside his head and also overtly through intense organized stalking. Many cannot function under this type of powerful psychological electronic attacks and in hope for a relief and in trust, fostered by the psychiatric community try medication which is ineffective because the targeted is technological

Six months into his job as a prosecutor, Myron May stated he just could not function and resigned. The technological effort had achieved a major part of the overall goal with Myron May and was on its way to fulfilling the rest of the plan for his usefulness. There are numerous horror stories within the targeted individual community of similar destruction of lives for many and varied reasons, the result the same.

In hope for relief, May sought the help of a psychologist. This in and of itself reveals someone trying to make sense of what was happening in his life and get coping help and at least a support system. A doctrine of this profession is "Crazy people don't think they are crazy."

As also mentioned earlier, patented EEG cloning systems allow downloading of a targets emotions, panic, fear, and anger, compatible as the target's biometric signature because it is of the target which is then beamed back to the target at the right moment for exploitation.

Make no mistake about it, this technology in research, TESTING, and development centering around Operation Paperclip scientific endeavors today, for well over 50 years, is no less than brilliant, very powerful, and, understand that testing is ongoing and widespread again justified and legalized after 9/11. Understand also that it is in the hands of twisted individuals, or so call patriots to Americans whose real patriotism is to globalization with no regard for anyone.

As I worked upstairs diligently on this book, the satellite was also making noises as if someone had entered my home downstairs and was walking around while continuously beaming panic and fear at me and while 100% beaming subliminal death threats as the thought of murder materialized in my mind. Regarding the walking sound downstairs this was amusing considering my home is carpeted.

May emerged from the first psychologist appointment with prescription for an antidepressant and an attention deficit drug, which he took faithfully until, about three weeks later; until suffered a panic attack at work.

After another attack, May returned to the doctor and was prescribed combination of Wellbutrin and Vyvanse. These are drugs which are documented to not cause paranoia although some media reports are now reporting they are.

Seeking help from law enforcement in these cases is fruitless. Many targeted individuals have had to accept this fact which closes the door for hope for any type of moral relief from covert terrorism. May actually told his friends that the officers at the Las Cruces Police Department laughed at him when he showed up on the morning of Sept. 7 to make the police report. The fact is it is a documented fact that law enforcement does set up secret/covert surveillances as a matter of course and is a documented fact and bug homes and vehicles. However, this reality pales in comparison to highly publicized mental illness as the source and technology cover-up.

In a similar case, on August 7, 2013, Aaron Alexis told Rhone Island police, he heard what he believed to be two black males and black female talking to him through a wall, but never saw them. He said he went to a hotel on the Navy base where he heard the same voices talking to him through the walls, floor, and ceiling. Alexis told police that the individuals were using "some sort of microwave machine" to send vibrations through the ceiling, penetrating his body so he couldn't fall asleep. In Alexis's case, he worked closely with DOD agencies within the Military Industrial Complex who have full knowledge of these types of devices and systems. When the police asked Alexis "… what the individuals were saying to him he would not elaborate."

Alexis "... is worried that these individuals are going to harm him." Alexis was also being threatened. Fear is a powerful motivator when a person is back against a wall or pushed over the edge.

Alexis was definitely officially military basically trained, but it doesn't appear that he received any extensive training in firearms or combat as shown by what appears to be confusion to how the weapon his is photographed as holding is used. Nonetheless, in both cases, the perceived, projected, conjured and targeted enemies were initiatively and primarily "Black Males" whose voices could be heard coming from and through inanimate objects. I think that African American men working these programs are strategically placed around other blacks so that racism cannot be used as a reason for the attacks by other races. Make no mistake about, in most cases; these individuals are acting under supervisory approval for what they are doing.

"It is my well-researched opinion of many that the Mark of the Beast, as related in scripture, is absolutely literal. Some believe that soon, all people on earth will be eventually coerced into accepting a Mark in their right hand or forehead for medical reasons, etc. Many reports this as an injectable passive RFID, radio frequency identification signal, transponder with a computer chip — a literal injection with a literal electronic biochip "mark" Could an implanted identification mark literally become Satan's Mark of the Beast ..." writes Terry Cook, The Mark of the New World Order (1996), p. 587

HP contractor Aaron Alexis may have been injected with the Mark of Beast transponder chip that coupled his individual and unique extremely low frequency wave lengths to a central computer that was able to put voices in his head, mind control and dispose of him in an operational (Governmental Gun Control Project) false flag attack on civilians which many reports also includes hiring of crisis actors playing important roles for perception management.

For example, at first glance, the slaying of Jordan Davis was a random act of violence. Jordan Russell Davis and Michael David Dunn were totally unrelated individuals until you factor in a common thread that ties Jordan Davis, Michael Dunn and Aaron Alexis together, Hewlett Packard (HP).

HP was also the contractor employer of Michael David Dunn.

The shooting of Jordan Davis occurred on November 23, 2012, at a gas station in Jacksonville, Florida. Jordan Russell Davis, a 17-year-old African American high school student, was fatally shot by Michael David Dunn, a 45-year-old software developer from Brevard County who was visiting the city for a wedding. The incident began when Dunn asked Davis and his companions to turn down the loud music that was being played in the vehicle in which Davis was a passenger. After the jury was unable to return a unanimous verdict on a charge of first-degree murder, the judge declared a mistrial on that count. Dunn was convicted, however, on three counts of attempted second-degree murder for firing at three other teenagers who were with Davis and one count of firing into a vehicle. None of the other teenagers were injured.

In closing arguments for the first trial, the defense lawyer for Michael Dunn cited the language of Florida's stand-your-ground law. Despite this, others have claimed that Florida's Stand Your Ground law did not play a role in the initial mistrial.

Dunn faced up to 75 years in prison for the four counts on which he was already convicted. Dunn's retrial for first-degree murder began the week of September 22, 2014. He was found guilty October 1, 2014, and was sentenced to a mandatory sentence of life in prison with no chance of parole on October 17, 2014

HP said to be a high-level New World Order/Luciferian- Bohemian Corporation. HP is one of world's leading purveyors of the "Mark of the Beast"- Global and National Biometric ID-Authentication Technology.

HP is also one of world's leading researchers and purveyors of human/computer nanotech biochip and transponders. Below is an old image depicting:

The Targeting of Myron May

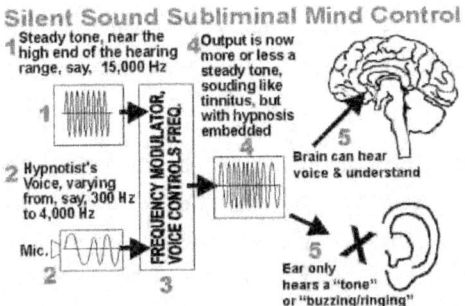

U.S. Patent 5159703 issued Oct 27, 1992

Myron May left the Police Department that day and went to a shooting range where friends had gathered for a bachelor party.

May, as stated in his story of "Being a Target" substantiates he went to the shooting range with associates. As they squeezed off rounds at the targets, May seemed agitated, they recalled.

He told them he wasn't sleeping because of his neighbors' constant spying which also combined noise. Their voices were keeping him up at night, he said. What he really wanted was to get a gun and take revenge on them. This statement could be interpreted as a suggestion via mind control efforts and strategic PSY OPS similar to the Facebook posting, of which no one would have a clue actually was connecting Myron May to a masterful set up "scheme" while setting the stage in the effort with full intent to push May into his purpose and over the edge in the "grand" scheme. Family and friends and Mr. Taunton's are on the record stating that May was not violent in any manner or even close to be. Again, the general consensus is that of a kind, gentle, humble, highly intelligent, sensitive soul.

In fact, in May's "To Do" list he lists the purchase of a weapon which apparently was not the weapon he used in the FSU shooting.

Because of these statements, supposedly, May's friends contacted his psychologist's office.

The psychologist made an appointment with May; May's friends reported, met with him for about an hour and then declared him to be fine. Shortly, after this episode, May reached out to a friend in law enforcement.

Technological perception management had May believing that everyone in the store where he was shopping had all eyes on him and were going to harm him. Many targeted individuals can give many examples of being somewhere and literally everyone around them being beamed suggestions making them useful. Many have not a clue from where a thought or action is materialized. This is another common trick reported by thousands of targeted individuals and of which Matt Barasch on the Dr. Phil show also tried to convey.

The fear frequency likely had Myron fearful for his life with valid reason it appears. No one in the store was after him with the exception of someone pulling his strings, technologically and using innocent people as pawns from miles away. Again, these people are having far too much fun in these operations destroying lives.

A day or two later, May again voluntarily checked himself into Mesilla Valley Hospital, a mental health center. He stayed for four days and on or about October 5, 2014, he abruptly drove nine hours to Denver, Colorado. He also told friends there were black cars following close behind. He then said that he would be a millionaire when he brought to justice the crooked cops who were persecuting him. Stopping for food or sleep was not an option, he said. He drove straight through the night.

After discharge from Mesilla, and during May's absence, his friends entered May's apartment to care for his dog. They conveniently found a prescription for Seroquel, an antipsychotic which he had not thought to take with him for his psychosis. After which his friends then contacted the hospital and were told there was nothing they could do and that May would have to come in on his own.

A valid question is does Seroquel work if you are not really psychotic and factually technologically targeted in what literally thousands of highly credible individuals continue to stand against and continue to reveal as being very real, factual, government human guinea technology testing program, sadly legal?

It was released that May had, keyword, **just recently** begun taking Seroquel. The generic name is Quetiapine. A research journal description of this drug associates it with violence as the promotion of mental illness continues to draw attention away from the cruelty and reality that Myron May was factually a Targeted Individual placed in a government very real program. When he began posting videos of this reality of thousands, this became an opportunity to portray how crazy conspiracy theorists really are. Information detailing the drugs violence intentionally published reads:

CONCLUSION

These data provide new evidence that acts of violence towards others are a genuine and serious adverse drug event that is associated with a relatively small group of drugs. Varenicline, which increases the availability of dopamine, and serotonin reuptake inhibitors were the most strongly and consistently implicated drugs. Prospective studies to evaluate systematically this side effect are needed to establish the incidence, confirm differences, among drugs and identify additional common features.

Get the picture?

This is the real hoax. Publishing of this medication prescribed to Myron May takes attention away from the reality of May's murder as a Targeted Individual created by the use of radio frequency technology influencing and physical torture on him and places the blame on medication given to a psychotic who just a few short months before had a history of credibility.

"On Oct. 7, two days after his trip to Colorado, May was driving the streets of Dona Ana County. He pulled into a sheriff's substation and dialed a name deleted number. He told her he couldn't take it; he was turning himself in.

As stated in his "Being a Target" letter, he then went to the police department desk to surrender, not once but three times, but was told he was not wanted on any charges. This was the result of the ongoing technological harassment to brainwash him. They were implanting that he was a drug addict, criminal, and by all accounts, vicious verbal abuse go hand in hand with these operations. Convincing the target that he is wrong in some way is brainwashing that legitimatizes these heinous manipulations for those doing the manipulations. Combined with everything else, and sleep deprived, May likely just wanted the severe psychophysical effort to stop.

After Myron May's murder, those responsible, in typical Nazi fashion continue to hide under the Nazi theme of "Just Following Orders."

"You can see the energy animating a person, by the love radiating out from their RIGHT eye. Left eye is WILL power. Now, look at the pics of Name Deleted and all these leaders. Always a lazy or dead right eye and an unusually larger left eye. That solar plexus and Sacral center energies dictating terms. Now, get any pics of people you KNOW to have good and loving hearts. Balanced eyes, very warm, and a very tight - not sagging - fluidity in their face. If they are being overshadowed - either good or ill - the overshadowing being's visage will show in their face. If it is heart center and higher, you'll see it instantly. A warm light about them. You know the rest..." CBS work and Rudolf Steiner

His right eye is dead. Has no life. The right eye radiates love and the left eye is the ray of will. In extremely advanced souls, both are in balance and highly radiant. Even someone lacking extra vision can easily see that. Now in darker souls and those on a satanic path, the right eye becomes lazy over the years, then finally "dies." After a bit of time, it doesn't even track on the same eye as their dominant left eye, which is will energy. And that fits, because, the satanic credo has but one law and one law only: Do what thou wilt shall be the whole law. And wilt, being old English for WILL. Ekhart Tolle by Don Bradley

A Monarch slave historically, as stated previously, was a person who had been put through a system of extreme trauma-based mind control in barbaric scientific studies. Today, these intense methods are no longer necessary due to radio frequency programming, resulting from years of testing. The objective then was to create mind-controlled Manchurians through splitting the walls of the human mind thereby to create multiple personality disorder or dissociative disorder. Each personality could then be programmed to perform specific tasks. Most personalities are unaware of each other and these "alters", as they are referred to, can be brought forward to take control of the body through the use of certain triggers. Each monarch slave has a handler.

Myron May alluded to the fact that he was aware that he had a handler who told him if he killed he would be free and released from prison of the "The Program" and destruction of his life. One of the signs of Monarch Programming, described in Monarch literature, as also stated previously is the droopy eye syndrome.

These images are listed only for consideration of the possibility and I cannot make claim to factual reality.

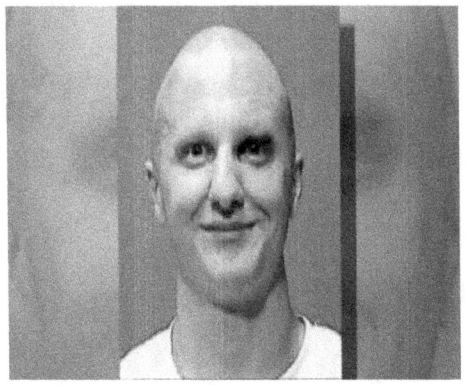

Jared Lee Loughner

Jared Lee Loughner pleaded guilty to 19 charges of murder and attempted murder in connection with the shooting near Tucson, Arizona, on January 8, 2011, in which he shot and severely injured U.S. Representative Gabrielle Gifford, his target, and killed six people, including Chief U.S. District Court Judge John Roll, as well as a 9-year-old bystander Christina-Taylor Green. Loughner shot and injured 13 other people, and one man was injured while subduing him.

Sanpaku is a term used to refer to a particular appearance of a person's eyes. Specifically, if the white of a person's eyes are visible either below or above the colored portion, or iris, that person is said to have Sanpaku eyes. This condition can either affect one or both eyes. It is said to reflect mental illness or many reports today likely Manchurian Delta Programming.

Dylan Klebold and Eric Harris

Eric Harris

Renee Pittman

Dylan Klebold

Aaron Alexis

Osama Bin Laden

Osama bin Laden, the founder and head of the Islamist militant group al-Qaeda, was killed in Pakistan on May 2, 2011, shortly after 1:00 am PKT. (20:00 UTC, May 1) by U.S. Navy SEALs of the U.S. Naval Special Warfare Development Group (also known as DEVGRU or SEAL Team Six). The operation, code-named Operation Neptune

Spear, was carried out in a Central Intelligence Agency-led operation. In addition to DEVGRU, participating units included the U.S. Army Special Operations Command's 160th Special Operations Aviation Regiment (Airborne) and CIA operatives. The raid on bin Laden's compound in Abbottabad, Pakistan, was launched from Afghanistan. After the raid, U.S. forces took bin Laden's body to Afghanistan for identification, then buried it at sea within 24 hours of his death, in accordance with Islamic tradition. The United States had direct evidence that Inter-Services Intelligence (ISI) chief, Lt. Gen. Ahmad Shuja Pasha, knew of bin Laden's presence in Abbottabad, Pakistan.

Seung-Hui Cho

Seung-Hui Cho was a South Korean mass murderer who killed 32 people and wounded 17 others on April 16, 2007, at Virginia Polytechnic Institute and State University in Blacksburg, Virginia. An additional six people were injured jumping from windows to escape. He was a senior-level undergraduate student at the university. The shooting rampage came to be known as the Virginia Tech shooting. Cho committed suicide after police breached the doors of the building where the majority of the shooting had taken place. His body is buried in Fairfax, Virginia.

Lee Boyd Malvo

Lee Boyd Malvo, also known as John Lee Malvo, is a Jamaican-American convicted murderer who, along with John Allen Muhammad, committed murders in connection with the Beltway sniper attacks in the Washington Metropolitan Area over a three-week period in October 2002. Although the pairing's actions were classified as psychopathy attributable to serial killer characteristics by the media, whether or not their psychopathy meets this classification or that of a spree killer is debated by researchers. In 2012, Malvo claimed that he was sexually abused by John Allen Muhammad.

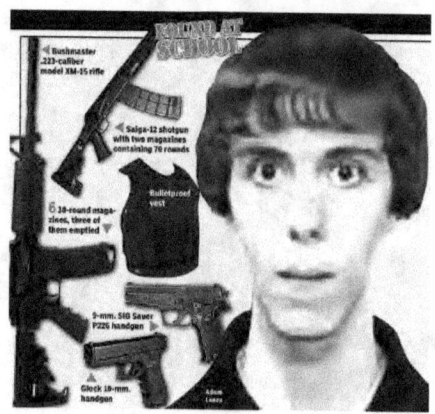

Adam Lanza

The Sandy Hook Elementary School shooting occurred on December 14, 2012, in Newtown, Connecticut, when 20-year-old Adam Lanza fatally shot 20 children and 6 adult staff members. Prior to driving to the school, Lanza shot and killed his mother at their Newtown home. As first responders arrived at the scene, Lanza committed suicide by shooting himself in the head.

Controversy still surrounds this incident as a hoax using crisis actors.

James Eagan Holmes (born December 13, 1987) is the accused perpetrator of a mass shooting that killed 12 people at a Century movie theater in Aurora, Colorado, on July 20, 2012

James Homes

James Holmes dyed his hair fire orange red as part of Batman Joker theme programming

Both images above show a child named Noah Pozner said to be murdered during the Sandy Hook mass shooting in December 14, 2012. The image below, December 19, 2014, show a Pakistan woman weeping for a boy said to be dead two years earlier.

Mohamed Atta

Atta was an Egyptian hijacker and one of the ringleaders of the September 11 attacks who served as the hijacker-pilot of American Airlines Flight 11, crashing the plane into the North Tower of the World Trade Center as part of the coordinated attacks. At 33 years of age, he is the oldest hijacker who had participated in the attacks.

Amanda Miller

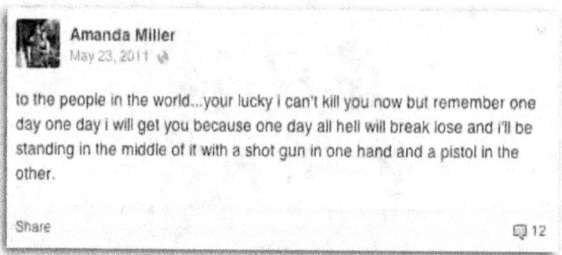

Is this another strategic forewarning?

placed strategically on Facebook?

The Targeting of Myron May

Jared and Amanda Miller Batman theme programming?

One half of the 2014 Las Vegas shootings team is Amanda Miller, occurring on June 8, 2014 in northeastern Las Vegas, Nevada by a married couple. Jerad and Amanda Miller, committed a shooting spree in which five people died, including themselves. The couple, who espoused extreme anti-government views, first killed two Las Vegas police officers at a restaurant before fleeing into a Walmart, where they killed an intervening armed civilian. The couple died after engaging responding officers in a shootout; police shot and killed Jerad, while Amanda committed suicide after being wounded.

Charles Manson

Years later proclaimed to be a victim of mind control

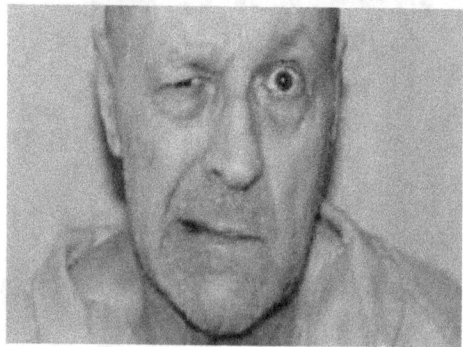

Robert B. Rhodes

SERIAL KILLER

Or perhaps, there is no connect at all and they all coincidentally suffer from Ptosis, which appears to be also known as Droopy Eyelid, and is common reported today. However, it is amazing that listed as one of many causes of Ptosis is factually Post Trauma.

Is this a coincidence or not? Frankly I don't know. Does the left eye droop denote loss of will over self, and the right loss of real love and compassion for other lives?

Christopher Dorner

Christopher Dorner, 33, an involuntarily terminated Los Angeles police officer, was named as a suspect wanted in connection with the series of shootings that killed four people and wounded three others.

Dorner spent 10 years in the U.S. Navy Reserves – rising to the rank of lieutenant before receiving an honorable discharge. He spent eight years at the Los Angeles Police Department, working as an officer in at least two divisions.

Cathy O'Brien

Cathleen Ann O'Brien proclaims she is a victim of the factual mind control government project named Project Monarch, which she said was part of the CIA's documented Project MKULTRA research, testing, and ultimately technology for behavioral engineering with the ideation for mass population mind control through various methods. O'Brien made these assertions in Trance Formation of America (1995) and Access Denied: For Reasons of National Security (2004) which she co-authored with her husband Mark Phillips.

"They were wrong. They never counted on the strength of the human spirit" ~ Cathy O´Brien

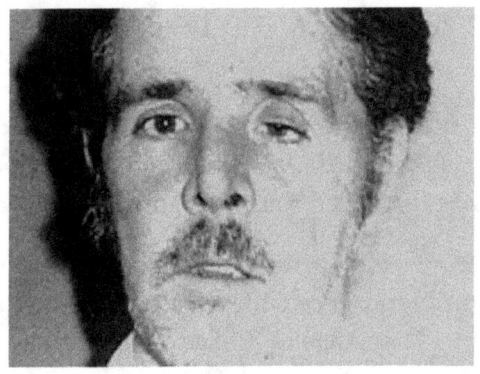

Henry Lee Lucas

SERIAL KILLER

The Targeting of Myron May

Charles Whitman

By all accounts, Charles Joseph Whitman, born June 24, 1941 had been brilliant with an IQ of 139, well balanced and a bright future. He had been an American engineering student and was former U.S. Marine before his life began to crumble. He killed 16 people and wounded 32 others in a spree shooting in Austin, Texas, on the University of Texas at Austin campus, in and around the Tower on the afternoon of August 1, 1966 when he snapped. Three people were shot and killed inside the university's tower and eleven others were murdered after Whitman fired at random from the 28th-floor observation deck of the Main Building. Whitman was shot and killed by Austin police officer Houston McCoy.

Prior to the shootings at the University of Texas at Austin, Whitman had murdered both his wife and mother in Austin. The shooting would remain the deadliest shooting on a U.S. college campus until the Virginia Tech shootings in 2007.

Charles Whitman stated that he wanted his brain examined after his death knowing something was wrong or controlling his will. It is documented that a tumor was located in his brain during autopsy, which today some believe, could have been swelling around an implant. Typically, people with brain tumors do not murder.

The Charles Whitman mass shooting would be the very first televised mass shooting and noteworthy, history documents, during a time when psychological electronic covert technology testing began intensely for mass population control due to global protests against war. The Arms Race ignited testing as far back as the 50's.

Samuel Holloway Bowers

Imperial Wizard of the White Knights of the KKK

Samuel Holloway Bowers was a leading white activist in the American state of Mississippi during the African-American Civil Rights Movement. In response to this movement, he cofounded a reactionary organization, the White Knights of the Ku Klux Klan which is still active today across the nation. Two notorious murders of civil rights activists in southern Mississippi are attributed to Bowers: the 1964 triple murder of Andrew Goodman, Michael Schwerner, and James Chaney near Philadelphia. for which he served six years in federal prison, and the 1966 murder of Vernon Dahmer in Hattiesburg, for which he was sentenced to life in prison 32 years after the crime. He also was accused of bombings of Jewish targets in the cities of Jackson and Meridian in 1967 and 1968 (according to the man who was convicted of some of the bombings, Thomas A. Tarrants III). He died in prison at the age of 82.

Essentially the Divide and Conquer agenda continues to thrive as a historical hope to separate humanity for NAZI Operation Paperclip ideology. There has always been an effort to destroy the human connection of all races necessary for continued control of the Elite who are the minority of 1%. If humanity united they would lose power due to colorful mixed family trees eradicating racism. Many may be used without consciousness of any possible motivating source after programming.

Some report that many have been programmed in America today. And as Jill Anjuli Hansen, coming from a family with close illuminati ties reported, the chip can be activated at any time and for any reason or purpose. Had Myron May's chip been activated?

However, the ultimate result of global efforts continues to be destruction of lives in many, many areas, to include ongoing war while advancing the agenda. And highly advanced technology today, highly perfected, is playing a key role which includes the patented weaponized "schizophrenia" technological effect.

CONCLUSION

A highly credible article released by Global Research, Professor James F. Tracy, entitled Florida State University Police Shot Attorney Myron May, Struck Nine Times in Hail of Gunfire, January 15, 2015, "Documents Reveals that officers Eschewed Non-lethal Force After Close-Range Encounter with FSU Gunman" The seven-page revealing article with other official documentation and images, in its entirety, can be located at the link as stated:

Excerpt below:

Police and forensic reports released this week to MHB by the Tallahassee City Attorney's Office indicate that five Florida State University and two Tallahassee police officers fired their .40 caliber handguns a total 34 times at 31-year-old FSU alum and Texas attorney Myron May on November 20, 2014. May, whose videotaped suicide note indicates he took up the mantle of mass shooter to draw attention to the plight of "targeted individuals", was struck nine times, with five of the officers' bullets hitting him in the rear of his pelvic area and lower extremities.

Additional document exhibits reveal that some witnesses, as shown in the article, attest that officers were within several feet of May prior to opening fire after he refused to drop his .380 caliber Lorcin pistol. It is also noted that responding Tallahassee and FSU police were both equipped with loaded Taser weapons which could have easily subdued May at such close range, yet instead "chose to dispatch May in a hail of gunfire."

All seven officers were exonerated by a Grand Jury in mid-December.

"The cops kept telling him to lay down to which the man made no moves," the signed testimony of one bystander reveals as shown in the image on the next page.

He kept his hands by his head, either scratching or resting them. I could not see if he still had his weapon draw[n] but upon the fifth request the cops unloaded about 15 rounds into him.

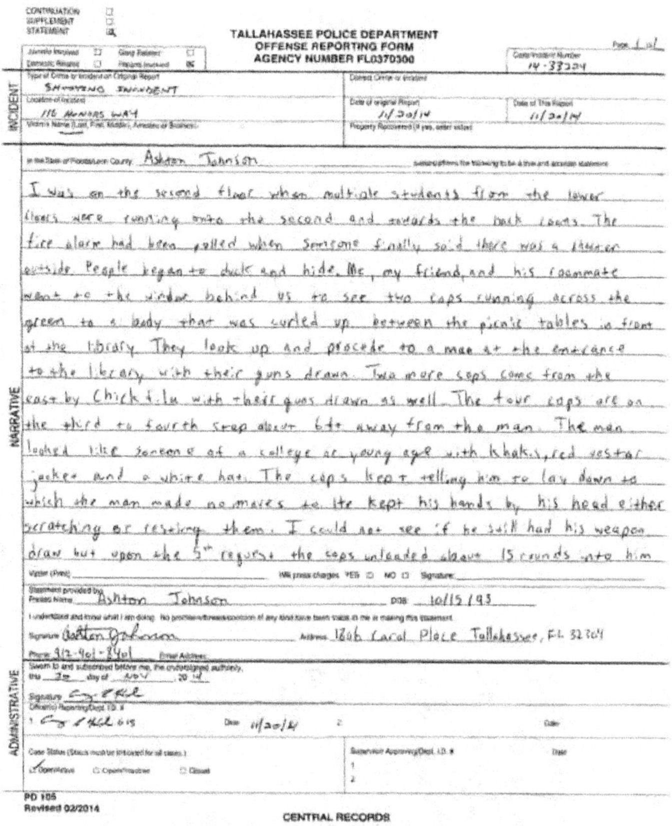

CONTINUATION ☐
SUPPLEMENT ☐
STATEMENT ☒

TALLAHASSEE POLICE DEPARTMENT
OFFENSE REPORTING FORM
AGENCY NUMBER FL0370300

Page 1 of 1

Case/Incident Number: 14-33224

INCIDENT

Type of Crime or Incident on Original Report: SHOOTING INCIDENT

Location of Incident: 116 HONORS WAY

Date of original Report: 11/20/14

Date of This Report: 11/20/14

NARRATIVE

In the State of Florida/Leon County: Ashton Johnson

I was on the second floor when multiple students from the lower floors were running onto the second and towards the back rooms. The fire alarm had been pulled when someone finally said there was a shooter outside. People began to duck and hide. Me, my friend, and his roommate went to the window behind us to see two cops running across the green to a body that was curled up between the picnic tables in front of the library. They look up and procede to a man at the entrance to the library with their guns drawn. Two more cops come from the east by Chick fila with their guns drawn as well. The four cops are on the third to fourth step about 6ft away from the man. The man looked like someone of a college or young age with khakis, red vest or jacket and a white hat. The cops kept telling him to lay down to which the man made no moves to. He kept his hands by his head either scratching or resting them. I could not see if he still had his weapon draw but upon the 5th request the cops unloaded about 15 rounds into him

Will press charges YES ☐ NO ☐ Signature:

Statement provided by Printed Name: Ashton Johnson DOB: 10/15/93

Signature: Ashton Johnson Address: 1806 Carol Place Tallahassee, FL 32304

Phone: 912-901-8401 Email Address:

ADMINISTRATIVE

Sworn to and subscribed before me, the undersigned authority, this 20 day of NOV, 20 14

Case Status: ☒ Open/Active ☐ Open/Inactive ☐ Closed

Supervisor Approving/Dept. I.D. # Date

PD 105
Revised 02/2014

CENTRAL RECORDS

Another Eyewitness Account by a Student and Police Report

Another excerpt from the Global Research article further states:

The Florida State University administration–overseen by powerful Florida GOP politico John Thrasher who was sworn in as FSU's president less than two weeks before the shooting–has repeatedly refused to release its police reports and related documentation of the incident. "We will be happy to provide records once they become public," Associate General Counsel Robyn Blank Jackson wrote in a January 9, 2015 email to MHB. "It is not possible to provide an exact date

Today, the reality of the hearing voices effect, is very real and substantiated by official patents, scientist, the Military Industrial Complex, etc., and powerful radio frequency, electromagnetic technology. As mentioned it goes by many names, and of which one is "Weaponized Schizophrenia." Today it is also being quite effectively used to officially discredit people, whistleblowers, activist, and human guinea pig test subjects, many of whom have held highly respectable and credible reputations prior to becoming a targeted individual. This capability is revealed also in the link on the subject of military weaponized schizophrenia shown excerpted below along with three examples from the link.

LINK:

http://ultraculture.org/blog/2014/06/06/world-military-weaponize-schizophrenia/

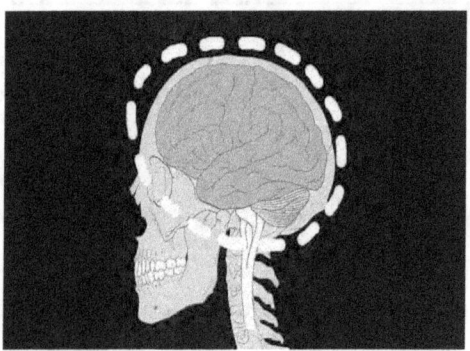

Here are three weapons technologies the US military has developed to directly monkey with people's nervous systems.

1. Active Denial System (ADS)—a.k.a. the "Pain Ray"

The Active Denial System (in its "non-lethal" iteration) can generate an electromagnetic energy burst that penetrates through clothing to the second layer of the skin, simulating the experience of opening a piping hot oven. The original purpose of the weapon was to incapacitate by way of raising a subject's body temperature—resulting in mental confusion, memory lapses and lethargy. Observed applications have been directed toward crowd control. Though non-lethal in its current state, the "Bio-effects" report kindly points out that "If the situation allows, and the source is sufficiently powerful, there is the possibility to use this technology in a lethal mode as well."

Even though the technology works, the weapon is too unwieldy for regular or spontaneous use. Shortfalls include a 16-hour start up time and diminished effectiveness during wet weather. A recent article from Correctional News reports that the Army continues to pursue ADS, demonstrating a smaller version for use at sea in September 2013. Raytheon, the firm that created ADS, is currently developing portable (possibly handheld) prototypes for use in law enforcement and corrections. Russia claims to have their version in an unknown stage of development.

(Below, check out an ADS demonstration from 2012.)

2. Microwave Hearing, Hypersonic and the "Voice of God"

Microwave hearing is a phenomenon that occurs when the brain is hit with an RF pulse. Every pulse that hits causes an expansion of the brain and the cochlea (inner ear auditory organ). This expansion results in a sound characterized as a buzzing or knocking inside the head. A successful experiment cited in the "Bioeffects" report includes the transmission of a voice counting from 1 to 10.

Taken from the report:

"Not only might it be disruptive to the sense of hearing, it could be psychologically devastating if one suddenly heard 'voices within one's head.'"

The Sierra Nevada Corporation was developing a weapon based on this concept called MEDUSA (Mob Excess Deterrent Using Silent Audio) to use in crowd control. The good news was that the tech would theoretically incapacitate crowds. The bad news was that they would have their brains cooked.

Alternative methods for creating a "Voice of God" weapon could come from Long Range Acoustic Devices (LRADS) or directed sound devices. LRADs are the very loud crowd control devices you can see being used on protesters, like at the Pittsburgh G20. Coincidentally, they are also used on pirates.

Portable versions of LRAD are being used from neighboring locations close to the target. It is 100% capable of beaming voices through the wall and is being used in ongoing technological harassment campaigns today by law enforcement and military personnel.

1. Friendly Japanese whalers give activists the finger while turning an LRAD on them. (Via Flickr)

An LRAD will cause permanent hearing damage to any unprotected individual standing within 100 meters. The military has also experimented with using spoken or whispered commands in an attempt to, I don't know, "Weapon: Death Ray Replaced by Voice of God."

Directed sound, like the sonic projector currently in development by DARPA, could be the next big step for this line of inquiry. Libraries, museums and even billboards are using commercial versions of the technology to provide audio only to those people standing in a narrow: "A Voice only you can hear: DARPA's Sonic Sound Weapon. "The video below shows one of these billboards in action. Listen for the whispering female voice as the camera approaches the billboard.

2. Neural Disruption

The military has looked at everything from strobe lights to Pokémon cartoons (no, they didn't say it by name, but what other Japanese cartoon has caused mass seizures?) to cause debilitating neural disruption in subjects. The MEDUSA weapon was designed with this purpose in mind, and we've since seen "Dazzlers" and "PHASR" technologies that use light pulses to disorient and stun their targets.

And the plot just thickened as this story was going to press—a "Brain circuit problem likely sets the state for the 'voices' that are symptoms of schizophrenics from St. Jude's Children's Hospital" has found that the auditory hallucinations or "voices" associated with schizophrenia could be linked to a chemical imbalance that interferes with communication between the thalamic inputs and auditory cortex, the sections of the brain that interpret audio stimuli.

The author of the study, Stanislav Zakharenko, M.D., Ph.D., was quoted as saying "we think that reducing the flow of information between these two brain structures that play a central role in processing auditory information sets the stage for stress or other factors to come along and trigger the 'voices' that are the most common psychotic symptom of schizophrenia..."

The study posits that the nerve pulses emanating from the auditory cortex are substantially weaker in schizophrenic patients than in "normal" individuals due to a genetic defect. New medications and treatments will be explored that can strengthen the signals from the firing neurons in the auditory cortex, resulting in the alleviation of systems.

Very possibly, of course, someone will take the research in a darker direction. (Is this a good time to mention that Putin's been talking about Russia working on Electromagnetic radiation gun. I'll be under my desk for a few minutes if anyone needs me.

(I, along with this author gratefully acknowledge the work of Sharon Weinberger, who has extensively covered the military applications of microwave weapons since 2007 for Wired.com and Nature.com of which this information was derived)

IS THIS ANOTHER CONSPIRACY THEORY BELOW?

Tunnel Boring Equipment

The Targeting of Myron May

Deep Underground Military Base Facility

NOTE: Above images said to be from an Anonymous hacker

FACT OR FICTION? YOU DECIDE…

Deep Underground Military Bases in America
(USED BY CONSENT of James Starfield, North Star Zone)

http://nstarzone.com/CODERED.html

Unknown to most Americans is a dark secret, and it's right under our noses. It's the reality of the existence of DEEP UNDERGROUND MILITARY BASES. These Underground bases get prominent play in dark rumors circulating about captured extraterrestrials and alien technology. The fringe culture rumors of underground alien-human shenanigans are in reality fed by leaks from questionable individuals, usually with intelligence connections. They are simply a ploy utility for the status quo. The whole captured-alien-hardware story is just a highly elaborate hoax to discredit those exposing the reality of these bases. They are also a cover for the wholesale looting of the federal treasury by the corrupt and cynical secret government. After Hurricane Katrina left the gulf coast region totally devastated, there were many witnesses coming forward with reports of UNMNTF and UNISF Troops working alongside the Army of the Republic of Mexico Soldiers in the New Orleans area.

When the day of Martial Law comes in America, the UNISF and UNMNTF troops located in Central America, in the US, and Canada will be deployed to help round up the millions of Americans whose names appear on the CIA Red List and the CIA Blue List. These troops are Chinese, Russian, German, Polish, Japanese, Ukrainian, Saudi Arabian, Pakistani, Mexican, Honduran, Salvadorian and Chilean, and many are stationed in the deep underground military bases. When that day comes in America, do not expect the Fox News Network, NBC, CBS, ABC, CNN, BBC News 24 or Reuters to give a full or accurate account of the truth. Mass detentions in camps and the underground bases, along with mass executions will occur, like they have in many other countries like Cambodia, Russia, China, Germany, Poland, Armenia, Georgia, Belorussia, Hungary and the Ukraine over the past 100 years. The best option for many Americans will be to have a safe place in a remote area where you can hide.

THE DENVER AIRPORT – Said to be a reprogramming site

The America you and your forefathers knew is coming to an end thanks to the Illuminati controlled secret government, and now they want America to become a Third World Nation ruled by a Fascist Police State, under their dictatorial control. This obviously cannot be achieved if America stays the way it is with many still being relatively well off, and still possessing firearms. Only the people of America can stop the coming American Holocaust from occurring. In America alone, there are over 120 Deep Underground Military Bases situated under most major cities, US AFBs, US Navy Bases and US Army Bases, as well as underneath FEMA Military Training Camps and DHS control centers. There are also many Deep Underground Military Bases under Canada. Almost all of these bases are over 2 miles underground and have diameters ranging from 10 miles up to 30 miles across!

The Targeting of Myron May

They have been building these bases day and night, unceasingly, since the 1940s. These bases are basically large cities underground connected by high-speed magneto-levity trains that have speeds up to 1500 MPH. Several books have been written about this activity. Richard Souder, a Ph.D. architect, risked his life by talking about this. He is the author of the book, 'Underground Bases and Tunnels: What is the Government Trying to Hide'. He worked with a number of government agencies on deep underground military bases. The average depth of these bases is over a mile, and they again are basically whole cities underground. They have nuclear powered laser drilling machines that can drill a tunnel seven miles long in one day. (Note: The September, 1983 issue of Omni (Pg. 80) has a color drawing of 'The Subterrene,' the Los Alamos nuclear-powered tunnel machine that burrows through the rock, deep underground, by heating whatever stone it encounters into molten rock.) The Black Projects sidestep the authority of Congress, which as we know is illegal. There is much hard evidence out there. Many will react with fear, terror and paranoia, but you must snap out of it and wake up from the brainwashing your media pumps into your heads all day long. Are you going to be a rabbit in the headlights, or are you going to stand up and say enough is enough? The US Government through the NSA, DOD, CIA, DIA, ATF, ONI, US Army, US Marine Corp, FEMA and the DHS has spent in excess of 12 trillion dollars building the massive, covert infrastructure for the coming One World Government and New World Religion over the past 40 years.

There is the Deep Underground Military Base underneath Denver International Airport, which is over 22 miles in diameter and goes down over 8 levels. It's no coincidence that the CIA relocated the headquarters of its domestic division, which is responsible for operations in the United States, from the CIA's Langley headquarters to Denver. Constructed in 1995, the government and politicians were hell bent on building this airport in spite of it ending up vastly over budget. Charges of corruption, constant construction company changes, and mass firings of teams once they had built a section of their work was reported so that no "one" group had any idea what the blueprint of the airport was. Not only did locals not want this airport built nor was it needed, but everything was done to make sure it was built despite that.

Masonic symbols and bizarre artwork of dead babies, burning cities and women in coffins comprise an extensive mural as well as a time capsule - none of which is featured in the airport's web site section detailing the unique artwork throughout the building. DIA serves as a cover for the vast underground facilities that were built there. There are reports of electronic/magnetic vibrations which make some people sick and cause headaches in others. There are acres of fenced-in areas which have barbed wire pointing into the area as if to keep things in, and small concrete stacks that resemble mini-cooling towers rise out of the acres of nowhere to apparently vent underground levels. The underground facility is 88.3 square miles deep. Basically, this Underground Base is 8 cities on top of each other! The holding capacity of such leviathanic bases is huge. These city-sized bases can hold millions and millions of people, whether they are mind controlled enslaved NWO World Army Soldiers or innocent and enslaved surface dwellers from the towns and cities of America and Canada.

There is Dulce Base, in New Mexico. Dulce is a small town in northern New Mexico, located above 7,000 feet on the Jicarilla Apache Indian Reservation. There is only one major motel and a few stores. It's not a resort town and it is not bustling with activity. But Dulce has a deep, dark secret. The secret is harbored deep below the brush of Archuleta Mesa. Function: Research of mind related functions, genetic experiments, mind control training and reprograming. There are over 3000 real-time video cameras throughout the complex at high-security locations (entrances and exits). There are over 100 secret exits near and around Dulce. Many around Archuleta Mesa, others to the south around Dulce Lake and even as far east as Lindrith.

Deep sections of the complex connect into natural cavern systems.

Level 1 - garage for street maintenance.

Level 2 - garage for trains, shuttles, tunnel-boring machines and disc maintenance.

Level 3 - everyone is weighed, in the nude, and then given a jump suit uniform. The weight of the person is put on a computer I.D. card each day. Change in over three dollars requires a physical exam and X-ray.

Level 4 - Human research in 'paranormal' areas - mental telepathy, mind control, hypnosis, remote viewing, astral traveling - etc. The technology is apparently here to allow them to know how to manipulate the 'Bioplasmic Body' Development of a laser weapon that can remotely cause burns and discomfort on its target. They can lower your heartbeat with Deep Sleep 'Delta Waves,' induce a static shock, then reprogram, Via a Brain- Computer link.

Level 5 -security is severe. Armed guards patrol constantly, and in addition to weight sensitive areas there (are) hand print and eye print stations. Here, is the device that powers the transfer of atoms.

Level 6 - Level 6 is privately called 'Nightmare Hall'. It holds the genetic labs. Experiments done on fish, seals, birds, and mice that are vastly altered from their original forms.

Then there is the Greenbrier Facility, in White Sulpfer Springs, West Virginia under the Greenbriar Resort. The Continuity of Government facility intended since 1962 to house the United States Congress, code-named Casper, is located on the grounds of the prestigious Greenbrier resort. The bunker is beneath the West Virginia wing, which includes a complete medical clinic. Construction of the facility, which began in 1959, required 2.5 years and 50,000 tons of concrete. The steel-reinforced concrete walls of the bunker, which is 20 feet below ground, are 2 feet thick. The facility includes separate chambers for the House of Representatives and the Senate, as well as a larger room for joint sessions. These are located in the "Exhibit Hall" of the West Virginia Wing, which includes vehicular and pedestrian entrances which can be quickly sealed by blast doors. They don't even hide this one, and it's even a tourist attraction.

The underground vault was built to meet the needs of a Congress-in-hiding - in fact the hotel is a replica of the White House. The underground area has a chamber for the Senate, a chamber for the House and a massive hall for joint sessions. Although the hotel says it gives tours of the 112,000 square area daily, the installation still stands at the ready, its operators still working undercover at the hotel. The secrecy that has surrounded the site has shielded it both from public scrutiny and official reassessment.

Most Americans will not believe that an American Holocaust will happen until they see it happening with their own eyes. Till then, it is just another strange conspiracy theory for them to laugh at. This is no laughing matter. When it happens, it will be too late to stop it. The US Government has been involved covertly in the creation of an army of loyal, brainwashed soldiers of the future. They will have cybernetic and microchip implants and will fight anywhere in the world, without question, with total loyalty and without hesitation or fear. These soldiers were created at Brookhaven National Laboratories BNL, the National Ordinance Laboratories NOL and the Massachusetts Institute of Technology MIT, and covertly transferred under DOD and NSA control and planning. Many of these soldiers are stationed in the Deep Underground Military Bases like the one under Denver International Airport. All of this information has been researched, and it has taken much effort to fit it together properly. There are many mag-lev subterranean train networks that stretch from these complexes and go out to other underground bases. All soldiers working in these bases are micro chipped and under total Psychotronic Mind Control.

Of the missing "Milk Carton People" that the FBI used to post on milk cartons, some were taken to these underground bases for genetic experimentation, micro chipping, psychotronic mind control and cybernetic implantations for future use as brainwashed soldiers of the

NWO. Every year in America hundreds of thousands of people go missing. The creation of a total Global Fascist Police State by the Illuminati will happen if we do not all wake up and see what is happening. I find it amazing that so many Americans, Scandinavians and Western Europeans refuse to believe that there are millions of UNISF and UNMNTF Troops in America. Under the Partnerships for Peace Program PFPP set up by President Bill Clinton in early 1993, thousands of troops a month have been coming into America. These Fascist criminals' parade as our friends and leaders, while stripping away democratic rights that will be replaced with a Corporatist and Fascist dictatorship, unless people, and especially Americans, wake up now. Here are the locations of some Deep Underground Military Bases in America:

ALASKA

1. Brooks Range, Alaska
2. Delta Junction, Alaska

    2a. Fort Greeley, Alaska. In the same Delta Junction area.

ARIZONA

1. Arizona (Mountains) (not on map) Function: Genetic work. Multiple levels
2. Fort Huachuca, Arizona (also reported detainment camp) Function: NSA Facility
3. Luke Air Force Base
4. Page, Arizona Tunnels to: Area 51, Nevada Dulce base, New Mexico
5. Sedona, Arizona (also reported detainment camp) Notes: Located under the Enchantment Resort in Boynton Canyon.

There have been many reports by people in recent years of "increased military presence and activity" in the area.

6. Wikieup, Arizona Tunnels to: Area 51

7. Yucca Mountain, Arizona

CALIFORNIA

1. 29 Palms, California Tunnels to: Chocolate Mts., Fort Irwin, California (possibly one more site due west a few miles)

2. Benicia, California

3. Catalina Island, California Tunnels to: I was told by someone who worked at the Port Hueneme Naval Weapons Division Base in Oxnard that they have heard and it is 'common rumor' that there is a tunnel from the base to this Island, and also to Edwards Air Force Base, possibly utilizing old mines.

4. China Lake Naval Weapons Testing Center

5. Chocolate Mountains, California Tunnels to: Fort Irwin, California

6. Death Valley, California

    Function: The entrance to the Death Valley Tunnel is in the Panamint Mountains down on the lower edge of the range near Wingate Pass, in the bottom of an abandoned mine shaft. The bottom of the shaft opens into an extensive tunnel system

7. Deep Springs, California Tunnels to: Death Valley, Mercury, NV, Salt Lake City

8. Edwards AFB, California Function: Aircraft Development - antigravity research and vehicle development Levels: Multiple Tunnels to: Catalina Island Fort Irwin, California Vandenberg AFB, California Notes: Delta Hanger - North Base, Edwards AFB, Ca. Haystack Butte - Edwards, AFB, Ca.

9. Fort Irwin, California (also reported detainment camp) Tunnels to: 29 Palms, California Area 51, Nevada Edwards AFB. California Mt. Shasta, California

10. Helendale, California Function: Special Aircraft Facility Helendale has an extensive railway/shipping system through it from the Union Pacific days which runs in from Salt Lake City, Denver, Omaha, Los Angeles and Chicago

11. Lancaster, California Function: New Aircraft design, anti-gravity engineering, Stealth craft and testing

    Levels: 42 Tunnels To: Edwards A.F.B., Palmdale

12. Lawrence-Livermore International Labs, California the lab has a Human Genome Mapping project on chromosome #19 and a newly built $1.2 billion laser facility

13. Moreno Valley, California Function unknown

14. Mt. Lassen, California Tunnels to: Probably connects to the Mt. Shasta main tunnel.

15. Mt. Shasta. Function: Genetic experiments, magnetic advance, space and beam weaponry. Levels: 5 Tunnels to: Ft. Irwin, California North

16. Napa, California Functions: Direct Satellite Communications, Laser Communications. Continuation of Government site. Levels: Multi-level Tunnels to: Unknown Notes: Located on Oakville Grade, Napa County, Ca. 87 Acres

17. Needles, California Function unknown

18. Palmdale, California Function: New Aircraft Design, Anti-gravity research

19. Tehachapi Facility (Northrop, California - Tejon Ranch Function:

Levels: 42 Tunnels to: Edwards, Llona and other local areas
Notes: 25 miles NW of Lancaster California, in the Tehachapi mountains.

20. Ukiah, California Function unknown.

COLORADO

1. Near Boulder, Co. in the mountains Function unknown

2. Cheyenne Mountain -Norad -Colorado Springs, Colorado Function: Early Warning systems - missile defense systems - Space tracking Levels: Multiple Tunnels to: Colorado Springs, Function: Early warning systems, military strategy, satellite operations

   Levels: Multiple NORAD is a massive self-sustaining 'city' built inside the mountain Tunnels to: Creede, Denver, Dulce Base, Kinsley.

3. Creede, Colorado Function unknown Tunnels to: Colorado Springs, Colorado - Delta, Colorado - Dulce Base, New Mexico

4. Delta, Colorado Function unknown Tunnels to: Creede Salt Lake, Utah

5. Denver International Airport (also a detainment camp) Function: Military research, construction, detainment camp facilities

   Levels: 7 reported Tunnels to: Denver proper, Colorado and Rocky Mountain "safe housing", Colorado Springs, Colorado (Cheyenne Mtn.)

6. Falcon Air Force Base, Falcon, Colorado Function: SDI, Satellite Control Levels: Multiple Tunnels to: Colorado Springs, possibly more.

7. Fort Collins, Colorado Function: Suspect high precision equipment manufacturing for space.

8. Grand Mesa, Colorado Function unknown

9. Gore Range Near Lake, west of Denver, Co. Function: Library and Central Data Bank

10. San Juan Valley, Colorado Hidden beneath and in an operating Buffalo Ranch Function unknown

11. Telluride, Colorado Function unknown

12. University of Denver, Co (Boulder area) Function: Genetics, geology/mining as related to tunneling and underground construction.

13. Warden Valley West of Fort Collins, CO Function Unknown Tunnels to: Montana

GEORGIA

Dobbins Air Force Base, Marrietta GA Function: test site for plasma and antigravity air craft, experimental crafts and weapons.

INDIANA

Kokomo, Indiana Function Unknown Notes: for years people in that area have reported a "hum" that has been so constant that some have been forced to move and it has made many others sick. It seems to come from underground, and "research" has turned up nothing although it was suggested by someone that massive underground tunneling and excavating is going on, using naturally occurring caverns, to make an underground containment and storage facility.

KANSAS

1. Hutchinson, Kansas Function unknown Tunnels to: Kinsley, Nebraska. Note: I received this report concerning this base: "I can vouch for the underground base in Hutchinson, Ks. The entrance to the tunnel is underneath Hutchinson Hospital and is huge. I was walking down that tunnel when I was a kid and at the end of this tunnel is a rock face wall with a 90 degree turn to the right. I stopped and refused to go any further down the tunnel, my instincts told me to stop. I distinctly remember hearing screams coming from there. These were not just cries for help, but more like blood curtailing horrifying screams like someone had just been murdered. I really want to know what was and probably still is going on down there. You are free to report the story, but please keep my name anonymous."

2. Kansas City, Kansas Function unknown Notes: Entrance near Worlds of Fun

3. Kinsley, Kansas Function unknown Tunnels to: Colorado Springs, Colorado; Hutchinson, Kansas; Tulsa Kokoweef Peak, SW California Notes: Gold stored in huge cavern, blasted shut. Known as the "midway city" because it's located halfway between New York and San Francisco.

MARYLAND

Edgewood Arsenal, Maryland (from Don) Martins AFB, Aberdeen Proving Ground, Maryland

MASSACHUSETTS

Maynard MA, FEMA regional center. Wackenhut is here too.

MONTANA

Bozeman, Mont. Function: Genetics

NEVADA

Area 51 - Groom Lake - Dreamland - Nellis Air Force Base Area 51 was said to exist only in our imaginations until Russian satellite photos were leaked to US sources and it's amazing how you can get photos all over of it now, even posters. They've been busy little bees building this base up.

Function: Stealth and cloaking Aircraft research & development. 'Dreamland (Data Repository Establishment and Maintenance Land)

1. Elmint (Electromagnetic Intelligence), Biological weapons research and genetic manipulation/warfare storage, Cold Empire, EVA, Program HIS (Hybrid Intelligence System), BW/CW; IRIS (Infrared Intruder Systems), Security: Above ground cameras, underground pressure sensors, ground and air patrol

2. Blue Diamond, Nevada Function unknown

3. Fallon Air Force Base area (the flats, near Reno) "American City" restricted military sites southwest of Fallon

4. Mercury, Nevada Function unknown 5. Tonopah, Nevada Function unknown 69: San Gabriel (mountains) On Western side of Mojave Desert Function unknown Notes: Heavy vibrations coming from under the forest floor which sounds like geared machinery. These vibrations and sounds are the same as heard in Kokomo, Indiana and are suspected underground building/tunneling operations.

NEW MEXICO

1. Albuquerque, New Mexico (AFB) Function unknown Levels: Multiple Tunnels to: Carlsbad, New Mexico Los Alamos, New Mexico Possible connections to Datil, and other points.

2. Carlsbad, New Mexico Functions: Underground Nuclear Testing Tunnels to: Fort Stockton, Texas. Roswell

3. Cordova, New Mexico Function unknown

4. Datil, New Mexico Function unknown Tunnels to: Dulce Base

5. Dulce Base, New Mexico. Tunnels to: Colorado Springs, Colorado Creed, Colorado Datil, N.M. Los Alamos. Page, Arizona Sandia Base Taos, NM

6. Los Alamos, New Mexico Functions: Psychotronic Research, Psychotronic Weapons Levels: Multiple Tunnels to: ALB AFB, New Mexico Dulce, New Mexico Connections to Datil, Taos

7. Sandia Base, New Mexico Functions: Research in Electrical/magnetic Phenomena Levels: Multiple Tunnels to: Dulce Base Notes: Related Projects are studied at Sandia Base by 'The Jason Group' (of 55 Scientists). They have secretly harnessed the 'Dark Side of Technology' and hidden the beneficial technology from the public.

8. Sunspot, NM Function unknown

9. Taos, New Mexico Function unknown Tunnels to: Dulce, New Mexico; Cog, Colorado Notes: Several other sidelines to area where Uranium is mined or processed.

10. White Sands, NM Function: Missile testing/design Levels: Seven known

NEW HAMPSHIRE

There may be as many as three underground installations in New Hampshire's hills, according to reports.

NEW YORK

New York, New York Function unknown Tunnels to: Capitol Building, D.C.

OHIO

Wright-Patterson Air Force Base - Dayton, Ohio Function: Air Force Repository. It is rumored to house stealth technology and prototype craft.

OREGON

1. Cave Junction, Oregon Function: Suspected Underground UFO Base Levels: At least one Notes: Suspected location is in or near Hope Mountain. Near Applegate Lake, Oregon, just over into California. Multiple shafts, access areas to over 1500 feet depth. Built using abandoned mine with over 36 known miles of tunnels, shafts.

2. Crater Lake, Oregon Tunnels: possible to Cave Junction

3. Klamath Falls, Oregon 4. Wimer, Oregon (Ashland Mt. area) Function: Underground Chemical Storage

   Levels: At least one

PENNSYLVANIA

Raven Rock, Pa (near Ligonier) Function: working back up underground Pentagon - sister site of Mt. Weather Notes: 650' below summit, 4 entrances.

TEXAS

1. Calvert, Texas Function unknown

2. Fort Hood, Texas (also reported detainment camp) Levels: Multiple

3. Fort Stockton, Texas Function: Unknown Tunnels to: Carlsbad, New Mexico

UTAH

1. Dugway, Utah Function: Chemical Storage, Radiation storage.

2. Salt Lake City Mormon Caverns Function: Religions archives storage. Levels: Multiple Tunnels to: Delta, Colorado & Riverton, Wyoming

VIRGINIA

Mount Poney - Near Culpepper, Virginia Function unknown

WASHINGTON

1. Mt. Rainier, Washington Function unknown. Levels: Multiple Tunnels to: Unknown Yakima Indian Reservation Function unknown Notes: Southeast of Tacoma Washington, on the Reservation, in an area 40 by 70 miles. Unusual sounds from underground (Toppenish Ridge). Low flying Silver Cigar shaped craft seen to disappear into the Middle fork area of Toppenish creek.

WASHINGTON DC

The Function: Part of a massive underground relocation system to house select government and military personnel in the event of cataclysmic event. Tunnels to: New York City; Mt. Weather.

WEST VIRGINIA

Greenbrier Facility, White Sulfur Springs, West Virginia under the Greenbriar Resort.

WYOMING: Riverton, Wyoming Function unknown Tunnels to: Salt Lake, Utah Denver, Colorado.

Again, fact or fiction, believe it or not, you decide...

One thing is certain. A human guinea pig technology testing program dating back decades is very real and in full force globally!

Stay Awake!

ABOUT THE AUTHOR

As I wrote this book, the operation center began to focus the beam on my eye along with intensified microwave deterioration of fluid in my shoulders and knees. Many targets report technological cataract from the vicious individuals working these centers and also joint deterioration which can lead to joint autoimmune disease.

After documenting, Ptosis, aka Droopy Eye Syndrome, the operation center began directing powerful, sporadic, piercing directed energy shots to my left eye. I noticed this after listening to my talking with another target on the phone and our discussing any possible connection. In the background during the call, I heard operation center personnel stating there may be a connection. They make it a point to insure I know they are listening to my every call, etc. This was before another person also listening in, a supervisor, said, "That's not what that's from." After the focused attacks on my left eye, I later felt that perhaps the new reasoning was that if my left eye drooped, I would discredit myself as part of the ongoing Spy Ops around me. Make no mistake about it, there is an ongoing 100% decisive effort to discredit me anyway possible covertly and overtly.

If droopy eye syndrome documented to be part of trauma-based programming, Google Monarch Droopy Eye, is factual, you can bet that those working these operations, have a droopy right or either left eye and it is likely pronounced and extreme. The right indicating a cold-blooded nature and the left indicating loss of will. My eyes still remain balanced.

Because they don't want to take me out quickly and reveal themselves, because of book exposure, and my proactive efforts everywhere, there is also very light focus from the beam on my heart / chest area sporadically. Albeit, once in a while focus here, ultimately, inevitably, could lead to heart disease, aka "slow kill" the strategic key being undetectable.

DON'T BEND, DON'T BREAK!

The End or is it the Beginning of Awareness?

The problem I had then, and to this day, which ultimately became an alarm for me that something behind the scene was afoot was/is how in God's name can anyone believe, especially a highly educated attorney, that killing would get his life back as an Assistant DA and post in on social network. Where is the logic?

And, could this distorted perception explain two credible reports of May disappearing for a period of time. Secondly, where did the gun come from if he did not bring the one he purchased with him in Las Cruces with him revealed in his suicide notes? Was he duped to play out a deadly scenario?

The fact is, several people during that timeframe also believe it was a set-up for me as well in an ongoing effort to discredit me, and many others, using, again, a "Kill two birds with one stone" psychological operation.

FOR THE RECORD

If YOU HAVE ANY THOUGHTS OF HURTING SELF OR OTHERS, WHETHER SUBLIMINALLY

INFLUENCED OR OTHERWISE, DO NOT, I REPEAT, DO NOT, CALL ME!!!

www.ingramcontent.com/pod-product-compliance
Lightning Source LLC
Chambersburg PA
CBHW071804080526
44589CB00012B/683